The Engineer's Guide to Materials

K. T. Voisey

The Engineer's Guide to Materials

An Applications-Focused Introduction to Materials Science

Springer

K. T. Voisey
University of Nottingham
Nottingham, UK

ISBN 978-3-031-62939-6 ISBN 978-3-031-62937-2 (eBook)
https://doi.org/10.1007/978-3-031-62937-2

This Springer imprint is published by the registered company Springer Nature Switzerland AG
The registered company address is: Gewerbestrasse 11, 6330 Cham, Switzerland

Preface

This book on materials is not for materials purists, but rather for engineers who whilst they make use of materials do not consider this their main interest. There are many text books which go into detail about the underlying bonding and crystal structures of materials. These areas are certainly important, but are not of immediate interest to all. Based on the premise that engineers are interested in "things" this book is applications led: applications for the main engineering materials are presented and explained in relation to relevant materials properties, and then the underlying mechanisms which produce, and importantly allow those materials properties to be deliberately manipulated, are explained. For those of whom this triggers interest and fascination and want to find out more there are many signposts to further information.

Nottingham, UK

K. T. Voisey

Acknowledgements

Thanks are due to the many people who have supported me during the preparation of this book. That includes both direct input from colleagues and contacts who have supplied information, images and advice on draft chapters and friends and family who have kept me going, chivvied when needed, not chivvied me when needed and distracted me when needed. It is dangerous to write a list as some will be missed out.

Thanks are also due to all those who have taught me and enabled me to make a living from researching and teaching materials (which while it definitely has its frustrations is generally an interesting and fun life). This certainly includes the many students I have taught at the University of Nottingham. These mainly arrive to study engineering, and it is an enjoyable challenge to engage them with the world of materials.

Contents

Introduction 1

1.1 What Do Engineers Need to Know About Materials?

The short version of the answer to the question "What do engineers need to know about materials?", as suggested by one of my mechanical engineering colleagues, is:

Strong and cheap? → steel
Strong and lightweight? → titanium
Lightweight and easy to machine → aluminium
Really lightweight for specialised applications → carbon fibre

This is not in itself wrong, but it misses out so much. There is a lot that materials scientists and materials engineers can do to deliberately manipulate the properties and performance of materials. There are also numerous phenomena that materials encounter in service or during handling which can impact how they perform. Having a working knowledge of the underlying mechanisms enables engineers to get the most out of materials.

The answer to almost every question about materials is "it depends", the key is knowing what factors "it" depends on, this book gives insight into this.

Selecting the right materials for any application and ensuring that appropriate asset management strategies are put in place are key to the efficient material usage required for sustainable engineering. The wrong material in the wrong place leads to material wastage, increased costs and increased energy input as components and structures fail and need to be replaced. Better understanding of materials leads to better use of materials. The consequent extended component lifetimes, decreased waste and maintenance requirements result in more sustainable engineering. Better understanding of materials is required for improving sustainability. Specific sustainability issues and concerns related to the topics considered are highlighted at the start of each chapter.

K. T. Voisey, *The Engineer's Guide to Materials*,
https://doi.org/10.1007/978-3-031-62937-2_1

This book is intended to give engineers an insight into materials and to equip them to be able to communicate effectively with the materials specialists they work alongside. The book has been written assuming a technically literature reader such as someone who has been, or is, studying engineering at an undergraduate or postgraduate level or a non-materials specialist engineer or scientist.

The materials covered range from the familiar, such as steel and glass, to the exotic, including nickel based superalloys and advanced composite materials. It concentrates on the main engineering materials that are used to create macroscopic components and structures. Electronic materials and properties and functional materials are not included.

People have always wanted to go further and faster as cheaply as possible. This has made the transport sector a lead in driving materials innovation. The majority of applications referred to in this book are from the transport sector. This both reflects an area where materials are of great importance and puts the various concepts covered into context by using familiar, accessible, examples. In all transport industries the concept of specific properties is very important. Specific properties such as specific strength and specific stiffness are simply the strength and stiffness divided by the density respectively. The mass of components required to meet operating demands is directly proportional to the values of these specific properties. Lower mass is advantageous as it decreases the energy, and hence cost and fuel consumption, required to move the vehicle.

Applications are central to this book. The approach taken is to showcase how materials are used, relating the various applications to the relevant material properties. Underlying mechanisms and concepts are then explained, along with their limitations. For those interested in learning more about the underlying materials science, there are many pointers to sources of further information. There is also a dedicated chapter on dislocations. The main chapters can be read without this but reading it will allow greater insight to be gained into how metallic materials work.

The overall aim is that the reader obtains a clear understanding of why you need to know more than just a material's chemical composition to understand how it will perform in service.

1.2 Manipulating the Properties of Materials

An understanding of what can be done to manipulate materials properties is important, as is understanding which properties can be easily changed and which cannot. A key point is to understand the connection between properties: you can never just change one property in isolation: increased strength comes at the price of decreased ductility.

The properties of metallic materials can be changed by changing the alloying additions. However, it is also important to be aware that the thermal and mechanical history of metallic alloys affect their properties. As soon as a metallic material has two phases the internal structure can be deliberately modified to control material properties. Four chapters are dedicated to applications of specific metallic materials and explaining the mechanisms used to control their properties: Chap. 2 focusses on aluminium alloys, Chap. 3 on titanium alloys, Chap. 4 on nickel based superalloys and Chap. 5 on steel.

For polymers, arranging the same molecule in different ways can change properties. Molecular chain length is also important. In both cases you need to know more than just the material composition to know what the overall properties and behaviour will be. Chapter 6 covers applications and the underlying material science of polymers and polymer composites.

Materials engineering can also transform ceramic materials from their everyday brittle, fracture prone form into state of the art materials that can withstand the aggressive operating conditions within gas turbine engines. Ceramics, along with the other brittle materials of concrete and glass are considered in Chap. 7.

Composite materials have their properties designed in during manufacture. Here it is important to understand the limitations between theory and reality. Polymer composites are included in Chap. 6 and ceramic matrix composites in Chap. 7.

1.3 The Reality of Materials in Service

An important part of materials science is understanding what happens to materials when they have been made into components and are put into service. Corrosion can be a big issue, as can wear. Creep and fatigue are phenomena that will be concerns under some conditions. These all progressively change the material with time, moving the properties and performance of the component away from the design specifications. Chapter 8 covers phenomena that can degrade materials in service and Chap. 9 describes approaches that can be used to protect materials and non-destructive testing methods which can determine the extent of material degradation.

1.4 Other Considerations in Material Selection

Material properties are clearly high up on the list when selecting materials for any application. However, there are numerous other considerations, not least cost. There can be many things contributing to cost: impact of geopolitical issues on material supply; inherent material cost; cost relating to complying with legislation; manufacturing cost. Chapter 10 introduces a systematic approach to ranking candidate materials.

As we push materials to be harder, stronger, tougher and able to withstand ever higher temperatures we are generating a series of new challenges for manufacturing. Strong, tough, high melting point nickel based superalloys are great for turbine blades but the same properties that make them so well suited to this application make them very difficult to process. Their high strength and toughness make conventional, subtractive, manufacturing processes more difficult. The high melting point means that any casting operations, where the molten alloy is shaped using moulds, need a material with an even higher melting point for the mould. Developments in manufacturing processes are driven by advances in materials to ensure that the best use can be made from materials. Chapter 11 highlights applications and properties of some additional materials as well as giving an indication of areas of current research in the extensive and rapidly moving area of materials science.

1.5 Is This Everything I Need to Know About Materials?

This book almost certainly does not give you absolutely everything you will need to know about materials. However, it will equip you with enough understanding to empower you to ask the right questions.

This book concentrates on engineering materials, that is materials used for macroscopic components. The final chapter has been used to showcase a range of other materials, including novel materials still being developed and specialised materials for niche applications. The range of materials is so large that no book is ever going to be exhaustive. This section aims to highlight how much more there is out there.

Suggestions for further reading, and how to keep up to date with the latest developments, are given at the end of each chapter. There are also sets of questions for each chapter so that you can check your understanding.

This text assumes a technically literate reader, so there is no explanation of concepts such as stress and strain. Relevant definitions are readily found in any standard engineering text book.

All chapters have references and a list of freely accessible resources where you can find out more about the particular topic. There are also many textbooks with titles such as "Introduction to Materials Science" where more can be found on specific aspects of materials.

The DoITPoMS resource hosted by The University of Cambridge has numerous tutorials which explain various aspects of materials: https://www.doitpoms.ac.uk/tlplib/phase-diagrams/index.php.

Aluminium Alloys

2

Aluminium alloys: the bullet points

- Vast majority of aluminium applications are of aluminium alloys, not the pure material
- Heating can change properties even if material is not melted, hence you need to know thermal history as well as composition to know the properties of an aluminium alloy
- Advantages: fairly corrosion resistant (but problems with salt), useful for minimising mass due to good specific properties
- Disadvantages: cost, maximum usage temperature is considerably less than melting point, welding issues

2.1 Introduction

Aluminium alloys are extensively used in aerospace. This is due to their good specific properties: their impressive strength to weight ratios mean that lower mass components can be made from aluminium alloys compared to other materials. This is good news for aerospace companies as lighter components mean lower fuel costs and more economic flying. Alternatively, this can be viewed as wasting less fuel, and money, on moving the aircraft itself and more on whatever profitable payload needs to be moved. However, aluminium alloys are not the cheapest materials. Another disadvantage is that though aluminium alloys can broadly be considered as corrosion resistant, and are certainly more corrosion resistant than steel, they can still experience corrosion, particularly in marine environments. Also, compared to other metallic materials they have a relatively low maximum operating temperature of about 200°C, though this does vary from alloy to alloy.

It is worth noting that the UK spelling is aluminium whereas the American spelling, aluminum, misses out that second "i", which changes the pronunciation.

K. T. Voisey, *The Engineer's Guide to Materials*,
https://doi.org/10.1007/978-3-031-62937-2_2

However you spell it, aluminium and aluminum are the same material. (Alumina, however, is the oxide, this is a ceramic and has very different properties to the metal!) The official, international standard, as set by the International Union of Pure and Applied Chemistry (IUPAC) is "aluminium". The words are so similar that the difference in spelling is easily overlooked. The main impact it has is that any web searches you do using one spelling will omit texts that use the other spelling, so you are advised to use both to avoid missing anything.

2.2 Sustainability Considerations Relating to Aluminium and Aluminium Alloys

Aluminium production is energy intensive, requiring a lot of energy to separate the metal from the ore. There are two important approaches to aluminium production which can make aluminium production "green". There is extensive usage of hydroelectricity, allowing aluminium production to take place with lower emissions [1, 2]. Well over a third, 39%, of worldwide aluminium production uses hydroelectricity. Recycling of aluminium waste also significantly contributes to improving the sustainability of aluminium. There is a dramatic energy saving, with recycling typically having 5% of the energy requirement of using new aluminium. This significant energy saving has led to aluminium being one of the most widely recycled materials, with more than 30 million tonnes of aluminium scrap recycled globally every year [3]. This makes aluminium alloys highly compatible with the principles of the circular economy, though alloy composition requirements do complicate this. Utilisation of aluminium alloys in the transport industries contributes to improving their overall sustainability as the lower mass aluminium alloy components result in lower fuel burn, or lower battery capacity, requirements, resulting in carbon emission and material usage savings respectively. Similarly, the use of aluminium alloys in electric vehicles decreases vehicle mass, increasing range between charges. Aluminium alloys are also used as sacrificial anodes to protect steel structures. Although the aluminium itself is consumed, the lifetime of the steel structure is extended due to the cathodic protection provided by the aluminium alloy anodes.

2.3 Applications of Aluminium

The good specific strength (strength to density ratio) of aluminium alloys makes them very attractive to all transport industries. There is extensive usage of aluminium alloys in aerospace, with aluminium alloys being the dominant material in civil aviation [4]. More than two thirds of Boeing's 747, 757, 767 and 777 aircraft structures are aluminium. Even the CFRP dominated Boeing 787 contains about 20% aluminium [5]. A similar pattern is seen in Airbus. The Airbus A320 has approximately 65% aluminium [6], the Airbus A380 has 61% aluminium, Airbus'

A350 XWB has 20% aluminium-lithium, an advanced alloy with a particularly low density [7].

The fuselage skin, wing skins, and airframe structural components of spars, ribs and stringers of civil aircraft are all frequently made from aluminium alloys (Figs. 2.1, 2.2 and 2.3). A variety of different alloys are used in aerospace, the majority being from the two and seven thousand series, with some usage of the six thousand series. The Boeing 737 fuselage skin is aluminium alloy AA 2024 [8]. The mixture of ductility, strength and fatigue resistance make this alloy well suited to the demands of this large, thin component that is subjected to repeated pressurisation cycles. The Boeing 777 also uses a two thousand series alloy for the fuselage body skin and AA2324-T39 for the lower wing skin surface, supported by AA2224-T3511 stringers. Seven thousand series alloys are used in areas which are more highly loaded. The Boeing 777 upper wing surfaces have AA7055-T7751 skins, supported by AA7075-T77511 stringers and AA7150-T77511 spar chords. AA7150 and AA7055 are also used in seat tracks, body stringers and body stiffeners. The Boeing 737 has AA7075 stringers and wing ribs. The higher strength of the seven thousand series is exploited here in these structural support components.

The Airbus A380 also makes significant use of aluminium alloys, with these forming 61% of the structure. Again, seven thousand series alloys are used for more highly loaded applications, with AA7449-T7651 being used for low-gauge wing ribs and AA7040-T7651 for the inner front and inner centre spars. The high fracture toughness of AA2024A-T351 is exploited in the lower wing skins and AA2024-T432 extrusions were used for fuselage frames.

In addition to the, post strengthening, final material properties that make these alloys well suited to their particular application the higher, pre-strengthening, ductility is also important. This allows the manufacturing processes of rolling to be used to produce the thin section material needed for wing and fuselage skins. This high ductility is also required in order to use extrusion, where the solid material is forced through a shaped die to generate shaped lengths with constant profile,

Fig. 2.1 Boeing 777 fuselage, made from 2000 series aluminium alloy. Photograph courtesy of Antony Smith

Fig. 2.2 Boeing 777 upper wing skin, made from 7055 aluminium alloy. Photograph courtesy of Ethan Smith

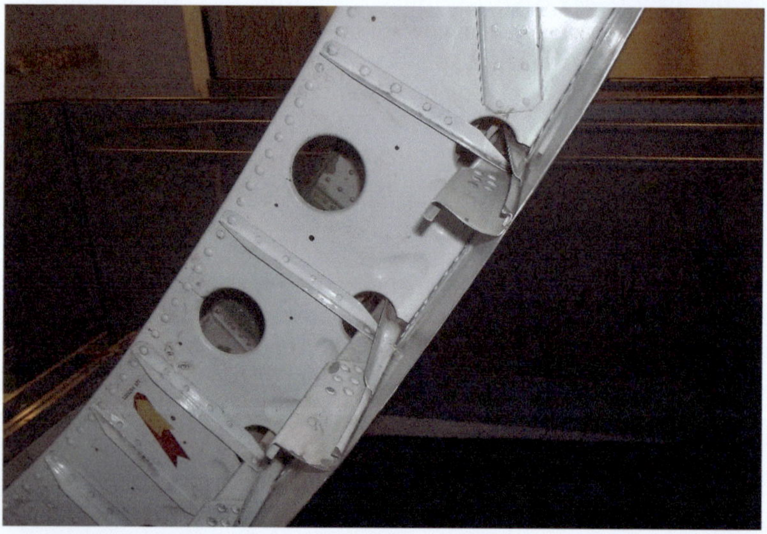

Fig. 2.3 Section through the fuselage of a Boeing 747 on display in The Science Museum, London. The outer skin is 2.2 mm thick, the skin, frame and supporting structures are all made from aluminium alloys. "Thin skin" by Elsie esq. is licensed under CC BY 2.0

such as are used in stringers and fuselage frames. In both cases the manufacturing processes would be far more difficult, and costly, without the ability to use heat treatment to switch between the ductile and high strength versions of the same alloy.

The six thousand series alloy AA6061 is known for its good corrosion resistance and weldability. The corrosion resistance and strength lead to it being used in helicopter rotor blades. It has various other aerospace applications including fuel tanks, and is also used in smaller, two seater, aircraft.

a

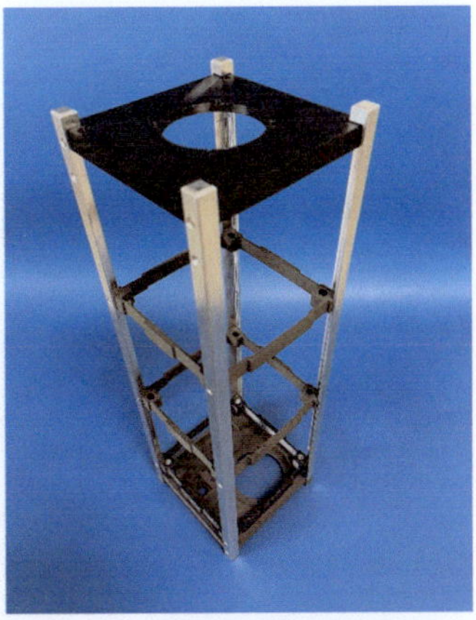

b

Fig. 2.4 CubeSat skeletons are made from aluminium alloys both CubeSats shown here have aluminium 6061 skeletons and windform XT, a carbon filled nylon, supporting structure. In each case the short side length is 10 cm. **a** Wormsail a two unit CubeSat and **b** skeleton of the AstroJam three unit CubeSat. Both satellites are from The University of Nottingham, images courtesy of Chantal Cappelletti

The space industry has particular focus on mass reduction, with it costing a reported £35k per kg put into orbit, spacecraft components often use aluminium alloys. CubeSats, of which well over 300 are now launched each year, use aluminium alloys for their main structures (Fig. 2.4). The alloys used vary from manufacturer to manufacturer and include AA6082, AA6061 and AA7075. The main material used to make the International Space Station is the aluminium alloy AA2219 T6.

SpaceX's Falcon 9 uses aluminium lithium, Al-Li, alloy tanks for its first stage [9] as well as for the second stage [10]. By alloying with lithium, the already low density of aluminium is pushed down below the usual 2700 kg/m^3 typical of aluminium alloys to lower values such as 2590 kg/m^3 and 2550 kg/m^3 for AA2090 (2.2%Li) and AA8090 (2.45%Li) respectively. The resultant mass savings on these rocket tanks translate into massive cash savings in the space industry. This makes it economically viable for the more expensive lithium alloys to be used here. Al-Li alloys are also now gaining ground, in the wider industry. There has been significant investment by Alcoa/Arconic in production of these alloys to feed demand from the aerospace industry. The Airbus A350 uses Al-Li alloys in the inner wing structure. Boeing have used AA2020 in the wings and horizontal stabiliser of its military aircraft the A5 Vigilante. NASA, SpaceX and Pratt & Whitney are among a number of other aerospace companies also finding commercial applications for these very low density alloys [11]. The third generation Al-Li alloys currently available have overcome the corrosion issues that were a feature of earlier versions.

More down to earth applications include growing use in shipping and rail, where the potential savings associated with mass reduction are driving a move away from

Fig. 2.5 A Deutsche Bahn ICE 3 high speed train, the 25 m long car bodies are made from aluminium alloy. "ICE 3 Fahlenbach" by Sebastian Terfloth User:Sese_Ingolstadt is licensed under CC BY-SA 2.5

steel. Railway car body applications exploit the good extrusion performance of the 6000 series, alloys such as AA6005 are exploited to form beams, sidewalls and the roof [12]. The ICE 3 (Fig. 2.5), RENFE Class 130/Talgo 250 and Shinkansen 300 series are amongst the various high speed trains where the car bodies are constructed from aluminium alloys in order to decrease weight [13, 14].

The good extrusion performance, corrosion resistance and weldability of AA6061 combine to make it well suited for use in bicycle frames. There is particularly extensive use of aluminium alloys in electric bike frames (Fig. 2.6). This is due to the wish to extend maximum range by minimising mass, important due to the need to counteract the added mass due to the motor and battery.

The automotive sector is still dominated by steel. However, there is growing interest in aluminium alloys. Overall, the usage varies considerably from manufacturer to manufacturer and from model to model. Audi has a particular interest in using aluminium, leading the mass market part of the sector in this area. The Audi A2 contained over 300 kg of aluminium, approximately a third of the vehicle mass and double the typical amount of 151 kg per vehicle [15]. Audi's space frame chassis, used in the Audi A8 (Fig. 2.7), combines aluminium castings, extrusions and panels. Continued extension of the use of aluminium, with some high strength steel, 9% of the body mass, has led to further weight savings. The body of the Audi A8 L is 241 kg [16]. In 2021, Nissan's shift to using lightweight aluminium alloy bonnet, doors and front wings saved 60 kg from the Qashqai body mass [17].

Jaguar use RC5754 [18], an alloy specifically developed for them. It contains up to 75% recycled material, which helps address the increasing demand for sustainable materials as well as delivering low mass components. The first application of RC5754 was in the Jaguar XE where it was used for body panels. Overall half of the body structure of the vehicle was made from aluminium alloy [19]. This is

Fig. 2.6 Electric bike with an AA 6061 frame. Image courtesy of Kay Bond

Fig. 2.7 An Audi A8, this vehicle contains for more aluminium than the average mass market car due to extensive use of aluminium in the space frame and body. "2012 Audi A8" by Brett Levin Photography is licensed under CC BY 2.0

one example of the electric vehicle sector's huge interest in low density materials such as aluminium as their priority is to extend the range and are willing to pay a premium to achieve this. Aluminium alloy tubing is being used in the cab frame of Ford's F-150 Lightning. Volkswagen's iD.3 has an aluminium alloy body and roof [20].

There are also many applications outside of transport. Extruded aluminium alloys are widely used in domestic window frames and doors as well as ladders and architectural trim (Fig. 2.8). The availability of interlocking, stiff and low weight, aluminium extrusions, also referred to as aluminium profiles, leads to widespread use in exhibition stands, shelving and partition systems (Fig. 2.8).

Aluminium's ductility is also exploited in applications such as drinks cans, food cans and cooking foil where it is the ductility which allows the material to

a b c

Fig. 2.8 Typical applications of aluminium alloys extrusions **a** poster frame, **b** shelving and **c** ladder be drawn, or rolled, out into thin sections (Fig. 2.9). For applications such as the gas cylinders shown in Fig. 2.10, which are made with AA6061 and AA7060, deformation to near final shape is done before precipitation strengthening heat treatments.

The high thermal and electrical conductivities of aluminium make it suitable for cookware and power lines respectively. Copper is a better electrical conductor than aluminium, however the significantly lower density of aluminium alloys, approximately a third of that of copper, more than compensates for this. Using aluminium alloys for overhead power lines gives lighter weight cables which need fewer pylons for support [21]. In this application a one thousand series alloy, usually AA1350, is used to carry the current. The one thousand series alloys are high purity

Fig. 2.9 The ductility and malleability of aluminium is demonstrated in its application in aluminium foil

a b

Fig. 2.10 The high ductility of aluminium alloys allows **a** billets to be drawn into extended geometries for application such as **b** gas cylinders.Images courtesy of Luxfer Gas Cylinders and hence have the higher electrical conductivity required by this application, however they are lower strength than other aluminium alloys. The required strength is provided by a core of a higher strength alloy such as AA6201 or steel [22].

Aluminium–silicon alloys have been widely exploited in casting applications. The reduction of melting temperature and viscosity means that these alloys have good mould filling characteristics, allowing detailed castings to be made, such as engine blocks and pistons. An aluminium alloy engine block represents a 50% weight saving compared to the competitor material of cast iron [23]. The high thermal conductivity and fatigue strength of aluminium alloys are exploited here, along with the good specific strength, resulting in the majority of modern engine blocks having been made from aluminium alloys (Fig. 2.11). Al-12Si is a common casting alloy as this composition has the lowest melting temperature. The four thousand series has silicon as the main alloying element. Al-Si alloys are used in a range of automotive castings where their good mould filling and low density combine to produce detailed castings with low mass. Ford have used an Al-Si-Mg alloy, equivalent to A356, for lower inlet manifolds [24].

Fig. 2.11 Aluminium alloys are used extensively in automotive engine blocks. "VW Lupo 1.4 16V Motor 05" by Mirko Junge is licensed under CC BY-SA 2.0

2.4 What Engineers Need to Know About Aluminium

2.4.1 Aluminium Alloys

Almost all engineering applications of aluminium use aluminium alloys as opposed to the pure metal.

Small amounts of alloying elements can have a large impact on the properties and performance of these alloys. The well known aerospace alloy, AA2024, has less than 10% in total of alloying elements but these can have a large impact, enabling yield strengths an order of magnitude higher than that of the pure metal. This also means that there can be small differences between the compositions of different alloys (Table 2.1). These do matter. Although the differences may appear small, they will contribute to impacting the overall properties.

There are many different aluminium alloys. The majority of these are wrought alloys which are usually referred to by 4 digit codes such as 2024, 7075 etc. Sometimes these are listed with a preceding AA: AA2024, AA7075. The AA stands for The Aluminum Association [25], an American body that registers and publishes specifications relating to aluminium alloys. Cast aluminium alloys are also classified by The Aluminum Association, these are again 4 digit codes but have a decimal point between the third and fourth digits, examples are 443.0 and 713.0. For both wrought and cast alloys the first digit reflects the main alloying element.

Wrought aluminium alloys are divided into series, such as the two thousand series and seven thousand series including 2024 and 7055. In any particular series all the 4 digit codes will start with the same number, all the two thousand series alloys, such as the widely used 2024, take the form 2xxx. Members of the same series share the same main alloying element, for the two thousand series that is copper, for the seven thousand series it is zinc. Alloys in the same series also share typical characteristics. The two thousand series alloys are relatively ductile whereas the seven thousand series alloys are known to be higher strength, leading to their use as aerospace stringers and wing ribs.

Table 2.1 Compositions of common aluminium alloys stated in weight %

	AA2024	AA6061	AA7055
Al, aluminium	93.5	97.9	87.4
Cu, copper	4.4	0.28	2.2
Mn, manganese	0.6		
Mg, magnesium	1.5	1.0	2.1
Si, silicon		0.6	
Cr, chromium		0.2	
Zn, zinc			8.2
Zr, zirconium			0.1

Looking up an alloy's code will let you find out the alloy composition but that is not all you need to know. The majority of wrought aluminium alloys are heat treatable. This means that their properties and performance depend not only on their composition but also on their thermal history. Alloy designations therefore frequently include information about what heat treatment has been performed. For 2024-T3 the "T3" is the heat treatment that been carried out and you can look up what that is, along with what properties the 2024 alloy will have after this heat treatment. Once an alloy has been bought in the composition is fixed. It is however possible to carry out additional heat treatment processes to change the properties of the material, AA2024-T3 can be changed to AA2024-T6 with appropriate heat treatment.

2.4.2 Need to Know Thermal History and Composition to Know Properties of Al Alloys

Many aluminium alloys are heat treatable. This means that any heat treatment, not only deliberate heat treatments, can change the properties and performance of aluminium alloys. Heat treatments that can change the properties of aluminium alloys do not have to melt them. This means that good records of thermal history have to be kept, and a good understanding of the impact of temperatures to be encountered in service is needed, and this will vary from alloy to alloy and will depend on previous thermal history. A detailed explanation of the key heat treatment steps is given below in the "How aluminium alloys work" section.

2.4.3 Corrosion Resistance

Aluminium alloys are relatively corrosion resistant, certainly compared to non-stainless steel. This is because the surface oxide layer which aluminium naturally forms when in contact with air is inherently protective. The surface layer of aluminium oxide (alumina) is thin, about 5 nm, but forms a good barrier to further oxidation as it is adherent, sealing off the underlying alloy from the environment. However, if this protective surface layer, the passive layer, is damaged corrosion can occur. One thing that can damage this surface layer is chloride ions, which are found in sea water, hence particular care needs to be taken in marine environments. In salt environments the chloride ions present destabilise the alumina layer, making it no longer protective and leaving the alloy prone to corrosion.

Corrosion of aluminium alloys can be exacerbated by galvanic corrosion between precipitates and the surrounding aluminium. However, this can be mitigated by use of Alclad. A number of alloys are also available in Alclad forms. This simply means that a thin outer layer of pure aluminium has been applied to enhance corrosion resistance. The layer is typically 20 μm in thickness so has a negligible impact on mass and the Alclad version of the alloy will have pretty much the same strength and stiffness as the non-Alclad version, just better corrosion resistance. The underlying alloy still benefits from the strengthening effects of precipitates formed due to alloying additions. However, because these are now sealed away from the surface by the pure Al layer, there is no danger of galvanic corrosion between the precipitates and the surrounding alloy.

Further protection by painting. Zinc chromate is commonly used in aerospace to provide corrosion resistance. The light green colour of zinc chromate means that many aluminium aerospace components appear light green in any images of aircraft construction. Zinc chromate is an example of a conversion coating. It enhances the formation of the surface oxide, converting some of the aluminium alloy into the oxide. The thicker oxide formed provides some increase in corrosion resistance but also offers an outer porous surface which enhances the adherence of paint. It is an effective undercoat.

Anodised aluminium is aluminium that has been electrically treated to enhance the growth of the surface oxide. The porous outer surface is good at picking up dye. The colourful wear resistance surfaces thus formed have many applications.

2.4.4 Disadvantages of Aluminium Alloys

The widespread usage of aluminium alloys is practical proof that their advantages outweigh their disadvantages, however there are some important disadvantages.

The first of these is cost. Aluminium is a highly reactive metal. This means it is tightly bound in its ore form, requiring significant energy input to extract the metal. Aluminium refineries hence tend to be located near sources of cheap electricity, such as hydroelectric facilities. Even so, the energy requirement pushes up the price of aluminium but also makes recycling of aluminium economically viable as this can save 95% of the energy required. Compared to other engineering alloys, aluminium alloys are typically more expensive than steels, but less expensive than titanium alloys. While the initial material cost may be high, using aluminium is often an overall economically viable choice. This is because the low density of the material, combined with the high strengths obtainable by the alloys, results in low mass components which decrease running costs as they require less energy, and hence fuel, to move.

Another disadvantage of aluminium alloys is their corrosion resistance. Ironically this can be regarded both as too good and not good enough (see Sect. 2.4.3 Corrosion resistance), depending on what aspect you are considering. The highly stable, high melting point alumina surface layer generated on aluminium alloys offers a fair degree of protection. However, for those applications where welding of aluminium is needed this oxide layer is problematic as it inhibits welding, meaning that TIG welding is generally used. This means that aluminium alloys cannot simply be swapped in for steel on manufacturing lines. Alternative joining techniques such as riveting and, more recently, friction stir welding can be used to join aluminium alloys.

2.4.4.1 How Aluminium Alloys Work—Precipitation Strengthening

Precipitate strengthening is what gives aluminium alloys their impressive properties. This only works in alloys. Combining the right alloying additions with appropriate heat treatment can increase the yield strength by approximately an order of magnitude compared to pure aluminium. What is happening is that the alloying additions allow the formation of small particles, precipitates, within the material. These precipitates are tiny but can be highly effective. The precipitates, only a few micrometres in size, are small particles, rich in alloying elements which disrupt the way the material deforms under applied stress. Different alloys have different precipitates forming. It is frequent to see significant changes in properties due to what may initially appear to be insignificant alloying additions: adding only 4.4% copper and 1.5% magnesium to pure aluminium is enough to transform it into AA2024, which dominates the aerospace industry.

The key thing that precipitates do is to make dislocation motion more difficult (see Chap. 12 on dislocations for more details). Since dislocation motion is required for the material to change shape, anything that makes dislocation motion more difficult will increase resistance to plastic deformation. This is the same as saying that the yield strength is increased.

Selecting the correct heat treatment for an aluminium alloy will determine both if any precipitates are present and how these are distributed. The same total volume fraction of precipitates will be more effective at strengthening if it is present as a fine distribution where there is a high number of small precipitates, as opposed to a smaller number of larger precipitates. The same alloy may be used in a number of different heat treated forms.

Precipitate strengthening does not work for every combination of alloying elements in aluminium alloys. A phase diagram is a really useful tool that can be used to identify if precipitate strengthening can occur. It can also be used to understand why the maximum operating temperatures of heat treated aluminium alloys can be considerably less than their melting points.

There is a lot of information in phase diagrams, and a lot of thermodynamics behind them. For the purposes of understanding aluminium alloys however, we can simply treat them as maps which show what phases are thermodynamically stable for any combination of composition and temperature. Remember that in materials the word phase is used to indicate a physically distinct form of matter, a broader definition than is used in physics or chemistry. In materials you can have multiple solid phases at the same time. A piece of cast iron with graphite flakes embedded in iron is described as a two phase material. Yes, both phases may be solid, that's ok, they are still physically distinct. When precipitates are present within a solid material that material is described as a two phase material.

The concept of inducing changes within a material without melting it is key to understanding how aluminium alloys work.

What is needed for precipitate strengthening to work is that the presence of the precipitates can be switched on and off by changing the temperature. This involves no change in overall composition of the alloy, which remains fixed. What is happening is that the alloying elements change between being spread out as individual ions and being clumped together in precipitates within the main material. Examination of a phase diagram will show if this is possible or not.

The key feature that is needed is a composition where the alloy exists in two phases at low temperature but where the same composition will be a single phase at higher temperature. This permits significant changes to the internal structure, the microstructure, of the material to occur. Precipitates can be made to appear, or disappear, by choosing appropriate heat treatments. And all this is done without melting the material.

Figure 2.12 is a schematic of the aluminium-copper phase diagram. This has the characteristics required for precipitate strengthening, specifically a single phase at high temperature and a two phase region at lower temperatures for the same composition. At the temperature—composition combination corresponding to point 1 the material will be a single phase. Alloying elements are present, but these will be distributed as individual atoms within the aluminium alloy. This kind of structure is referred to as a solid solution because the alloying elements are regarded as being dissolved in the aluminium. The use of the word "solution" here can cause confusion with non-metallurgists. The heat treatment processes described here all take place in the solid state, no melting of the material is required. Point two is also in a single phase field, but it is a different phase, in this case the phase referred to as θ. In both cases, the microstructure would only feature one type of material. This is as indicated in the schematic microstructures. The only features visible would be the grain boundaries, separating different crystals of the same material. Point three is different, it is in a two phase region. Now the microstructure features two different forms of material, both the main aluminium material, the matrix, and the embedded alloy element enriched precipitates, shown as red spots in the schematic microstructure.

The key feature of the Al Cu phase diagram in Fig. 2.12 is that the material can change from being a single phase material to a two phase material simply by changing the temperature. This enables the heat treatment processes describe below to occur. Points 1 and 3 are at the same point on the x axis, meaning that they both have the same overall composition. Figure 2.12 shows that the two phase structure at point 3 can be changed to the single phase structure of point 1 simply by changing the temperature. The overall composition is unchanged, what changes is whether or not the alloying elements cluster together to form the precipitates. This transformation is key to the sequence of heat treatment processes that underly precipitate strengthening.

Solutionising is the starting point for precipitate strengthening. Perhaps surprisingly, what solutionising does is to actually get rid of any pre-existing precipitates. The solutionising step increases the temperature of the alloy to move it into the single phase region (Fig. 2.13). This means that any precipitates will now dissolve. The composition doesn't change. The alloying elements are still present. What has happened is that now, instead of being clumped together to form precipitates, the alloying elements are dissolved in the main material and spread out as individual atoms/ions within the main crystal structure. An important point to

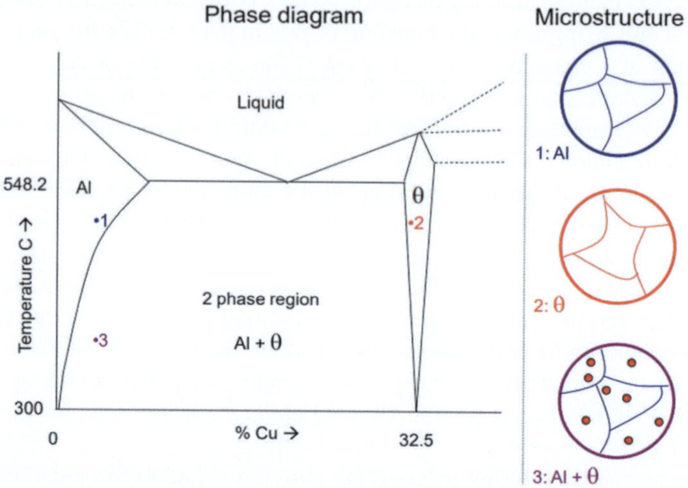

Fig. 2.12 Schematic of the Al-Cu phase diagram and corresponding microstructures for the highlighted points. The circular schematic microstructures are intended to indicate the view seen down a microscope in each case

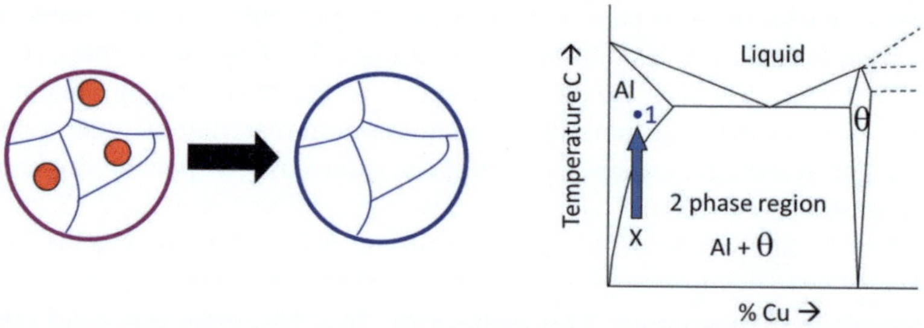

Fig. 2.13 Schematic showing the change in microstructure during solutionising, and the relevant part of the phase diagram

note is that whatever the initial state of the alloy, after the solutionising step the result is always the same: no precipitates. This state is also referred to as a solid solution. This indicates that alloying elements are present but they are dissolved in the main material. Solutionising can be regarded as transforming the alloy into a blank sheet of paper, generating a uniform starting point for any subsequent heat treatment, regardless of the prior history of the material.

Point 1 in Fig. 2.12 is a suitable temperature for solutionising of the given composition. It is not the only option, any other temperature within that single phase field, labelled as "Al" would also be a valid choice.

After solutionising, the material is precipitate free, however it is still at high temperature, within the single phase region. The next stage of heat treatment,

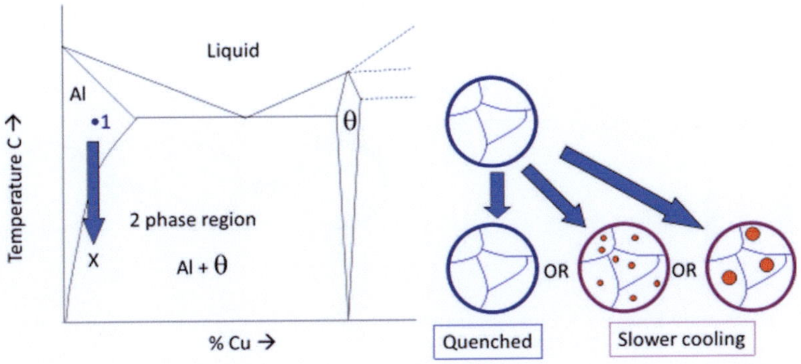

Fig. 2.14 Schematic illustrating quenching, and changes in microstructure if cooling is slower with relevant parts of phase diagram indicated

quenching, is really important, even though nothing happens. In fact, it is really important that nothing happens. In quenching the temperature is decreased very, very quickly. Looking at the phase diagram will now indicate that the material is within the two phase region. However, if quenching has been done successfully then the single phase, precipitate free, solid solution will have been retained (Fig. 2.14).

Quenching uses kinetics to cheat thermodynamics. The phase diagram shows the thermodynamically stable phases for any combination of temperature and composition. While phase diagrams show which changes will occur, they do not contain any information about how long it takes for changes to occur.

To understand this, it is useful to consider a drink with ice in it. It is thermodynamics which tells us that the ice will melt, but to understand how long it will take an understanding of kinetics, which includes how quickly atoms move around inside a material is needed. For precipitates to form, the alloying elements need to move within the material so that they clump together. This is achieved by diffusion. Basically, the atoms jump from site to site within the material, sticking together if they bump into other alloying elements. This process takes time and is highly temperature dependent. What quenching does is that it cools the material so quickly that there simply is not enough time for the alloying elements to diffuse. As diffusion is thermally activated, once the material has been cooled diffusion is not possible so the high temperature structure is frozen in, even though thermodynamics (the phase diagram) says it should no longer exist.

Commercially, quenching is achieved by dropping hot material out of a solutionising furnace straight into a water or oil bath. This is needed to get the high cooling rates required. Direct contact of the water or oil with the hot metal enhances the rate of cooling. The thicker a material section is the more challenging it is to achieve fast enough cooling in the centre of the material, the material which will cool most slowly.

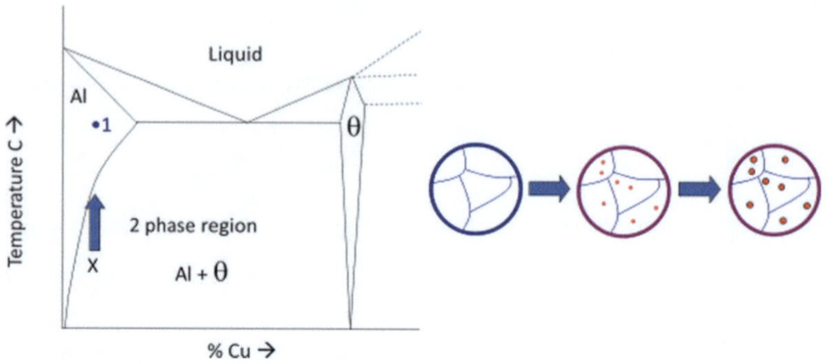

Fig. 2.15 Schematic highlighting relevant part of phase diagram and microstructural changes during ageing

The final part of the heat treatment process is ageing. This is when the precipitates form, and is why precipitate strengthening is also referred to as age hardening. In ageing the material is heated up, but this is done within the two phase region (Fig. 2.15). This provides thermal activation energy for diffusion, so the alloying elements are now mobile and can join to form precipitates.

Ageing itself can be broken down into three stages (Fig. 2.16).

Nucleation—this is when the system changes from having zero to some precipitates. Embryonic nuclei are constantly forming and redissolving, with only nuclei that reach a critical size being energetically favourable to survive and grow into precipitates.

Growth—nucleation has already occurred. Each viable nuclei now grows, accumulating additional alloying elements until all atoms of the alloying elements have joined a precipitate.

Coarsening—this is generally undesirable. During coarsening the amount of precipitate does not change, only the way it is distributed. If a fine precipitate distribution is held at temperature it will evolve to become a coarser distribution with fewer, larger precipitates (Fig. 2.17). What is happening is that the precipitate material is being redistributed via diffusion. This process is driven by a reduction in interfacial area. This is the same effect as surface tension driven coalescence of

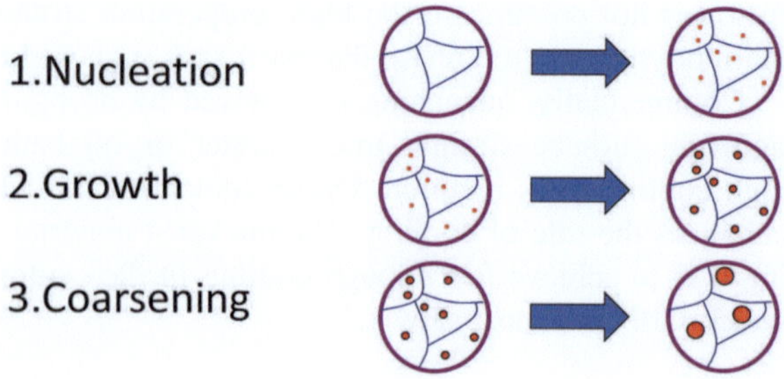

Fig. 2.16 Schematic showing the changes in microstructure during the three different stages of ageing

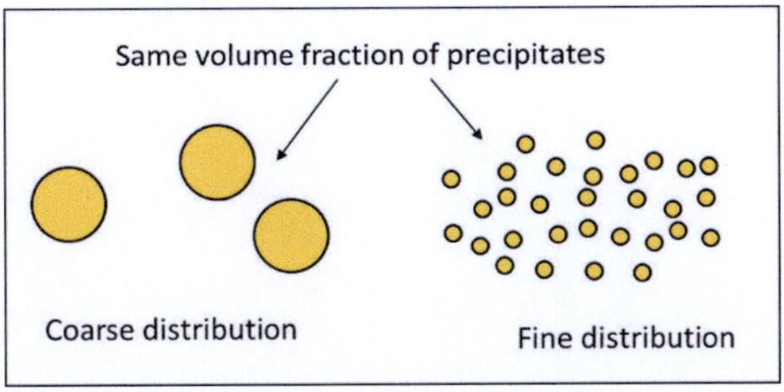

Fig. 2.17 Change in precipitate distribution due to coarsening

water droplets. As coarsening is thermally activated, it can be avoided by avoiding high temperatures.

Figure 2.18 schematically shows how the overall heat treatment and phase diagram relate to each other.

If coarsening is not avoided, it is seen that the strengthening effects of the precipitates start to be lost (Fig. 2.19). The force needed to move a dislocation through a field of precipitates is inversely proportional to the precipitate spacing. This is why yield strength decreases as coarsening progresses. As a precipitate distribution changes from a fine distribution, with a large number of small precipitates, to a coarser distribution with a smaller number of larger precipitates, the spacing between precipitates increases. It is the desire to avoid coarsening, and the associated loss of strength, that restricts the usage of aluminium alloys at high temperatures. Maximum operating temperatures do vary from alloy to alloy but are typically below 200 °C.

One strategy that can be used to extend the service life of aluminium alloys is to use the material in the slightly underaged form. This means that though the

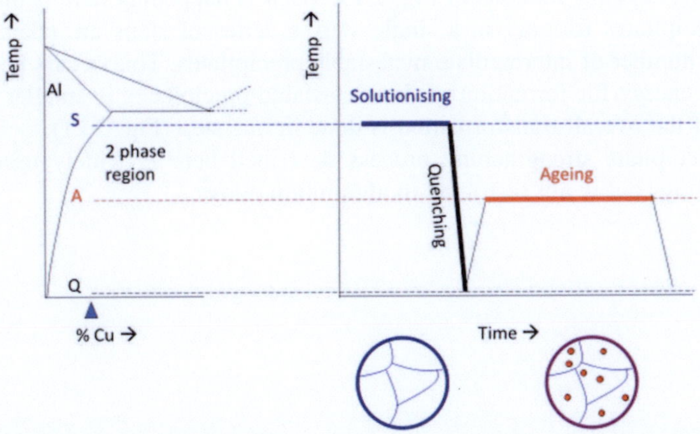

Fig. 2.18 Schematic heat treatment for precipitate strengthening of an Al-Cu alloy, with corresponding schematic phase diagram

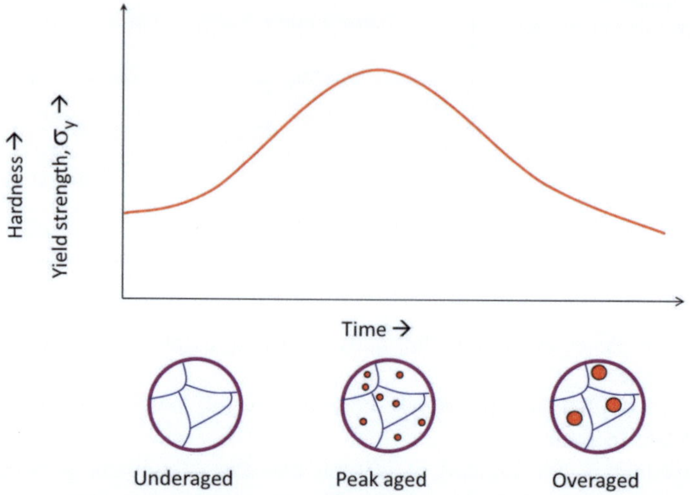

Fig. 2.19 Variation of properties during age hardening

alloy is not operating at maximum strength, any exposure to high temperature will first complete precipitation before the system moves on to coarsen. Strength will then initially increase before dropping off as usual. This strategy extends the operational lifetime before the detrimental effects of coarsening kick in, but this is at the expense of operating at sub-maximum material strength.

The consideration of precipitate strengthening described here produces the variation of strength seen in Fig. 2.19. A more realistic view is given in Fig. 2.20. The overall shape is similar, the differences are due to this overall behaviour being the combined effect of several different precipitates forming, the summation of several different overlapping versions of Fig. 2.19. What is happening here is that instead of the precipitates forming in a single step, a series of steps are taken, passing through a number of intermediate metastable precipitates. This occurs because the activation energy for formation of the metastable precipitates is smaller than that required if the overall transformation is done in one step (Fig. 2.21).

The precipitate strengthening process described here is widely used in aluminium alloys but is not restricted to aluminium alloys.

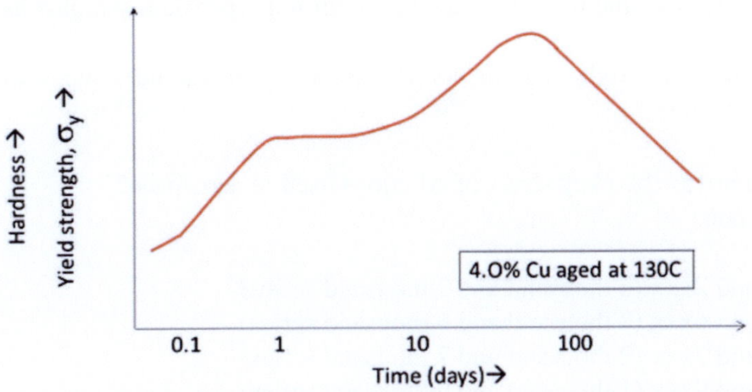

Fig. 2.20 Real variation of strength during precipitate strengthening and coarsening, after [26]

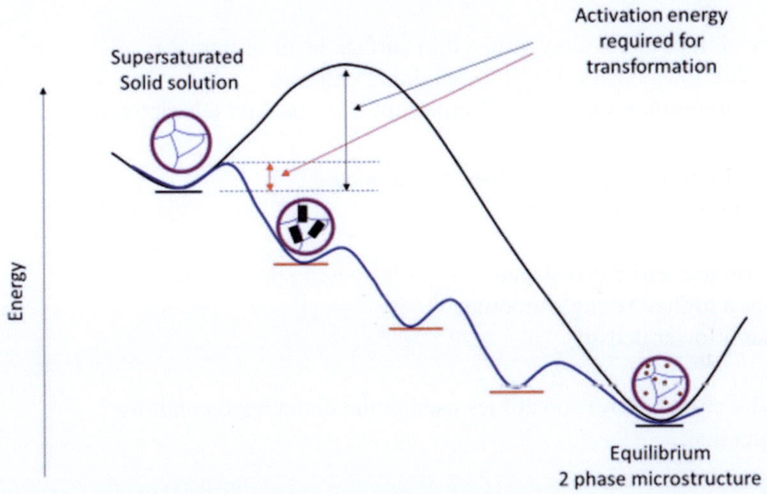

Fig. 2.21 Schematic comparison of the energy activation barriers with single step and multistep transformations

2.5 Test Your Understanding—Questions

Q1 You have two samples of the same alloy, you run multiple tests on these and it is clear that the two different samples have very different properties. You check the chemical composition and that is what is expected for your alloy for both samples. What is going on?

Select one:

a. the two samples have undergone different heat treatments and/or mechanical processing

b. this happens all the time, it is annoying but not a problem you just have to live with it
c. this cannot be right, surely some mistake, you should keep checking the measurements

Q2 What are the two key series of Al alloys used in aerospace?
Select one:

a. 6xxx and 2xxx (6 thousand and 2 thousand series)
b. 7xxx and 6xxx (7 thousand and 6 thousand series)
c. 2xxx and 7xxx (2 thousand and 7 thousand series)
d. 1xxx and 9xxx (1 thousand and 9 thousand series)

Q3 What is Alclad?
Select one:

a. A version of an Al alloy with a thin surface layer of pure Al
b. An ultra low density Al alloy containing chlorine
c. An aluminium trade body (ALuminium CLaims And Distribution)

Q4 What improved property does Alclad have?
Select one:

a. it is more corrosion resistant
b. it has a higher Young's modulus
c. it has a lower density

Q5 Why are Al alloys sometimes used in the under aged condition?
Select one:

a. It does not matter, it is the same alloy so the properties will be the same
b. This extends service life, initial strength increases before then decreasing with time
c. Under aged material has lower total precipitate volume so is less dense

Q6 Why are rivets, rather than welds, commonly used to join Al alloys?
Select one:

a. you can't get an airtight seal with a weld
b. rivets are cheaper
c. Al alloys are difficult to weld

Q7 Why are Al-Li, aluminium–lithium, alloys of interest?
Select one:

a. They have a much higher Young's modulus than other Al alloys
b. They are very cheap
c. They have an extra low density

Q8 You have bought in a precipitate hardened alloy but when you check it under a suitable microscope there are no precipitates, what has happened?
Select one:

a. the precipitates have probably fallen out during transport of the material, check the packaging carefully to find them
b. the wrong material has clearly been supplied, you need to contact your supplier and get the correct material
c. this is not necessarily a problem, the material has probably simply been supplied in the solutionised plus quenched state

Q9 What is the very fast cooling step in heat treatment called?
Select one:

a. coarsening
b. quenching
c. nucleation
d. solutionising

Q10 What is another term used for over ageing?
Select one:

a. winnowing
b. solutionising
c. coarsening

Q11 What feature of a phase diagram is needed for a material to be suitable for precipitate hardening?
Select one:

a. you must have two adjacent single phase fields
b. you must be able to transition from a two phase field to a single phase field by increasing the temperature
c. there must be no two phase fields

Q12 You are carrying out heat treatment to precipitate strengthen a 2xxx Al alloy, at the end of the process you realise that you have not achieved the desired precipitate distribution. Your precipitate distribution is coarser than you wanted. What affect does this have on the properties of your alloy?

Select one:

a. the yield strength will be lower than wanted
b. the Young's modulus will be lower than wanted
c. no change to properties, you have some precipitates so all is ok
d. the yield strength will be higher than expected

Q13 You are carrying out heat treatment to precipitate strengthen a 2xxx Al alloy, at the end of the process you realise that you have not achieved the desired precipitate distribution. Your precipitate distribution is coarser than you wanted. Is there anything you can do to get the desired precipitate distribution?

Select one:

a. it is ok, bigger precipitates are better, this is not going to be a problem
b. oh no! this is a disaster, you have to throw away the material, buy some more and start again
c. fear not, you can simply start again from the beginning

Q14 Precipitate strengthening is often used to increase the yield strength, but that is not the only property it changes, which of these properties will also change when precipitate strengthening is used?

Select one:

a. Young's modulus
b. density
c. hardness

Q15 Fill in the following table to ensure you understand what is happening at each stage of the heat treatment process.

	Phases	Number of ppts	Ppt size	Yield Strength
As received	?	?	?	?
Solutionising				
Quenching				
Ageing -nucleation				
-growth				
-coarsening				

Table Q15

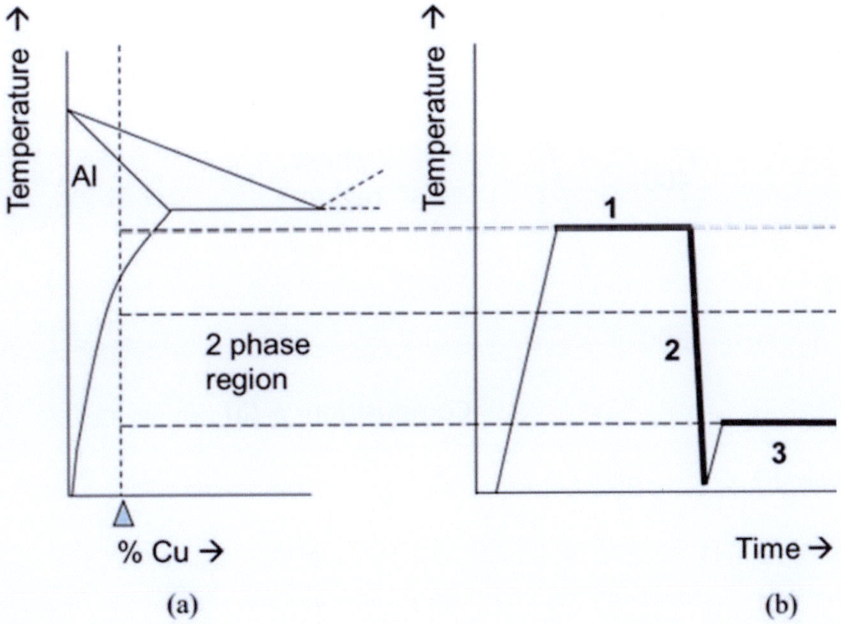

Figure Q16

Q16 Explain what happens in each of the stages 1, 2, and 3 in the figure above, for an alloy with the composition indicated by Δ. Your answer should include a diagram of the microstructure at the end of each stage.

Q17 Using the schematic phase diagram give guidance on what temperatures are suitable for solutionising and ageing of an alloy with the composition corresponding to the dotted line. What can you say about the maximum operating temperature of the alloy?

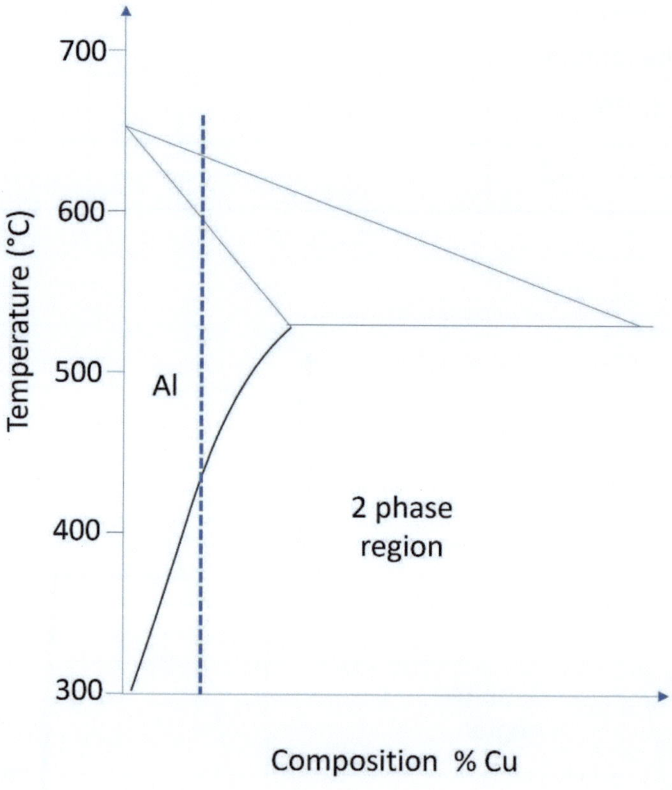

Figure Q17

2.6 Test Your Understanding—Answers

Q1 Answer a the two samples have undergone different heat treatments and/or mechanical processing.

Q2 Answer c 2xxx and 7xxx (2 thousand and 7 thousand series).

Q3 Answer a A version of an Al alloy with a thin surface layer of pure Al.

Q4 Answer a it is more corrosion resistant.

Q5 Answer b This extends service life, initial strength increases before then decreasing with time.

Q6 Answer c Al alloys are difficult to weld.

Q7 Answer c They have an extra low density.

Q8 Answer c this is not necessarily a problem, the material has probably simply been supplied in the solutionised plus quenched state.

Q9 Answer b quenching.

Q10 Answer c coarsening.

Q11 Answer b you must be able to transition from a two phase field to a single phase field by increasing the temperature.

Q12 Answer a the yield strength will be lower than wanted.

Q13 Answer c fear not, you can simply start again from the beginning.

Q14 Answer c hardness.

Q15 Answer

	Phases	Number of ppts	Ppt size	Yield Strength
As received	?	?	?	?
Solutionising	?→1	?→0	-	
Quenching	1→1	0→0	-	
Ageing -nucleation	1→2	↑	↑	↑
-growth	2	No change	↑	↑
-coarsening	2	↓	↑	↓

Q16 Answer

1. solutionising—any existing precipitates are dissolved, diagram should show single phase microstructure
2. quenching—rapid cooling, no change in microstructure, diagram should match that for 1
3. nucleation or first stage of ageing, nuclei form, diagram should show 2 phase structure with nuclei

Q17 Answer Solutionising needs to be in the single phase region, labelled Al, temperature should be above 430 °C and less than 600 °C, for ageing the temperature needs to be within the 2 phase region, temperatures up to 430 °C are acceptable, not higher.

The maximum operating temperature certainly needs to be less than 430 °C (otherwise the precipitates and associated strengthening will be lost) it should also be less than whatever temperature is stated as ageing temperature.

Further Reading

Organisations and companies

There are a number of organisations dedicated to aluminium, and which have informative websites containing a lot of information, resources and data relating to all aspects of the aluminium industry.

The International Aluminium Institute describes itself as the only body representing the global primary aluminium industry. Its website, https://international-aluminium.org/ has resources and information relating to all aspects of aluminium, including https://thealuminiumstory.com/ which is a good overview of aluminium.
The Aluminum Association represents aluminium production and jobs in the U.S.A. Its website https://www.aluminum.org/resource-library has many freely available resources on all aspects of aluminium.
Alfed is the Aluminium Federation, described as the voice of the UK aluminium industry https://alfed.org.uk/
European Aluminium aims to influence EU policy to support the aluminium industry. Its website, https://european-aluminium.eu/, houses numerous aluminium related resources and reports on various aspects of the industry.
A number of aluminium companies also have useful resources available via their websites: https://www.shapesbyhydro.com/en/material-science/, https://www.hydro.com/en/aluminium/industries/, https://clintonaluminum.com/

Open access articles and research papers

Aluminium Insider has news articles on all aspects of the industry https://aluminiuminsider.com/
US Navy's "Brand New" Aluminum Ship: Foiled by Seatwater, Mark Thompson, Time, July 05 2011, https://nation.time.com/2011/07/05/u-s-navys-brand-new-aluminum-ship-foiled-by-seawater/ this article highlights issues related to corrosion of aluminium alloys in marine environments.
The Use of Aluminium in the EV Market, August 17, 2022, Metals Warehouse https://www.metalswarehouse.co.uk/the-use-of-aluminium-in-the-ev-market/
Aluminium Usage in Cars Surges as Automotive Industry Shifts Towards Electrification Brussels, 2 May 2023, Press Release, European Aluminium. https://european-aluminium.eu/wp-content/uploads/2023/05/23-05-02-European-Aluminium_PR_Aluminium-Usage-in-Cars-Surges-as-Automotive-Industry-Shifts-Towards-Electrification.pdf
Aluminum Alloy Development for the Airbus A380, sep 2009, Total Materia https://www.totalmateria.com/page.aspx?ID=CheckArticle&site=ktn&NM=227
Applications of Aluminum Alloys in Rail Transportation, Xiaoguang Sun, Xiaohui Han, Chaofang Dong and Xiaogang Li https://doi.org/10.5772/intechopen.96442 in A. Dobrzański L (ed.) (2021) Advanced Aluminium Composites and Alloys. IntechOpen. Available at: https://doi.org/10.5772/intechopen.87723.
https://www.aircraftaluminium.com/application/202425247075-aircraft-aluminum-for-fuselage-skins.html

References

1. Harbor aluminium "Green Aluminum: The Sustainable Metal" accessed 18/11/23 https://www.harboraluminum.com/en/green-aluminum#:~:text=Environmentally%20friendly%20aluminum%20is%20considered,can%20be%20considered%20Green%20Aluminum.
2. https://primary.world-aluminium.org/aluminium-facts/energy-water/ accessed 18/11/23
3. https://international-aluminium.org/work_areas/recycling/#:~:text=Aluminium%20is%20one%20of%20the,recycled%20materials%20on%20the%20planet accessed 28/11/2023
4. Yue Wu 2022 J. Phys.: Conf. Ser. 2228 012024 Application of aluminum alloy in aircraft https://doi.org/10.1088/1742-6596/2228/1/012024
5. The New York Times July 29, 2013 The Jump to a Composite Plane
6. Dolganova, I., Bach, V., Rödl, A. et al. Assessment of Critical Resource Use in Aircraft Manufacturing. Circ.Econ.Sust. 2, 1193–1212 (2022). https://doi.org/10.1007/s43615-022-00157-x
7. G. Poulton, Aviation's material evolution, Airbus Newsroom 18 February 2017 https://www.airbus.com/en/newsroom
8. The Boeing 737 Technical Site, http://www.b737.org.uk/production.htm#:~:text=Wing%20upper%20skin%2C%20spars%20%26%20beams,compressive%20strength%20to%20weight%20ratio accessed 16/5/23
9. SpaceX https://www.spacex.com/vehicles/falcon-9/ accessed 29/7/24
10. SpaceX CRS-6 Mission Press Kit, April 2015 https://www.nasa.gov/sites/default/files/files/SpaceX_NASA_CRS-6_PressKit-2.pdf
11. G. Djukanovic Aluminium-Lithium Alloys Fight Back, Sep 5, 2017, Aluminium Insider https://aluminiuminsider.com/aluminium-lithium-alloys-fight-back/
12. Stephan W Kallee and Calvin Blignault, TWI Ltd, Friction Stir Welding for the Fabrication of Aluminium Rolling Stock, European Railway Review, Issue 3, May 2008 https://www.twi-global.com/technical-knowledge/published-papers/friction-stir-welding-for-the-fabrication-of-aluminium-rolling-stock-may-2008
13. The 10 fastest high-speed trains in Europe, Railway Technology, November 27, 2013 https://www.railway-technology.com/features/feature-the-10-fastest-high-speed-trains-in-europe/
14. X. Sun, X. Han, C. Dong, and X. Li, 'Applications of Aluminum Alloys in Rail Transportation', Aluminium Alloys. IntechOpen, May 31, 2021. https://www.intechopen.com/chapters/75650
15. Aluminum Content in Cars, Summary Report, Public Version, prepared for European Aluminium by Ducker Worldwide, June 2016 https://www.alfed.org.uk/files/Fact%20sheets/european-aluminium-ducker-study-summary-report_sept.pdf
16. https://www.audi-mediacenter.com/en/vorsprung-durch-technik-redefined-the-audi-a8-l-2745/body-2837 accessed 18/11/23
17. Lightweight aluminium gives Qashqai a cutting edge, Nissan Motor Corporation, Official Great Britain Newsroom, 2021/05/27 https://uk.nissannews.com/en-GB/releases/release-7aa4d3e4b25782dcbedfa7df60065421-lightweight-aluminium-gives-qashqai-a-cutting-edge?selectedTabId=releases
18. Novelis Works with Jaguar Land Rover to Create RC5754 High Recycled Content Aluminium Alloy for Automotive Industry PR Newswire, Atlanta, Feb 11, 2016 https://www.multivu.com/players/English/7755351-novelis-jaguar-rc5754-recycled-aluminum-alloy/ accessed 16/5/23
19. Jaguar Land Rover Upcycles Aluminium to Cut Carbon Emissions by a Quarter, JLR Newsroom, 21 August 2020 https://media.jaguarlandrover.com/news/2020/08/jaguar-land-rover-upcycles-aluminium-cut-carbon-emissions-quarter
20. https://www.car.info/en-se/volkswagen/id3/id3-2022-25247085/specs accessed 18/11/23
21. https://www.aluminiumleader.com/application/electrical_engineering/ accessed 18/11/23

22. Types of Conductors Used in Overhead Power Lines, K Daware, peak demand utility products https://peakdemand.com/types-of-conductors-used-in-overhead-power-lines/#:~:text=ASCR%20conductors%20are%20very%20widely%20used%20for%20all%20transmission%20and%20distribution%20purposes.&text=ACAR%20conductor%20is%20formed%20by,(6201%20aluminum%20alloy)%20core accessed 18/11/23
23. The Aluminium Automotive Manual, Chapter 2 Applications – Power Train – Engine Blocks, Version 2011, European aluminium Association https://european-aluminium.eu/wp-content/uploads/2022/11/aam-applications-power-train-2-engine-blocks.pdf
24. Dayalan Gunasegaram, D.J. Farnsworth and T.T. Nguyen, Identification of critical factors affecting shrinkage porosity in permanent mold casting using numerical simulations based on design of experiments, February 2009 Journal of Materials Processing Technology 209(3) https://doi.org/10.1016/j.jmatprotec.2008.03.044
25. The Aluminum Association, https://www.aluminum.org/resource-library accessed 18/11/23
26. After Porter & Easterling, Phase Transformations in Metals and Alloys, 2nd Ed, fig, 5.37, itself after Silcok, Heal & Hardy, Journal of the Institute of Materials vol 82 p239 (1953–1954).

Titanium Alloys 3

Titanium alloys: the bullet points

- Maximum operating temperatures are significantly less than melting points as heating can change properties, even if material is not melted, via solid state phase transformations
- Advantages: useful for minimising mass due to low density and good specific properties, good corrosion resistance
- Disadvantages: relatively high cost and poor tribological behaviour
- The two different phases alpha and beta result in three different alloy types: alpha, beta and alpha plus beta. Understanding the characteristics of the alpha and beta phases enables understanding of the different alloy families.
- Titanium alloys have fatigue limits and are not susceptible to fatigue if operated below their fatigue limit

3.1 Introduction

Titanium, or rather titanium alloys, are a premier engineering material. They have an attractive combination of low density and high strength. The correspondingly impressive specific properties mean that low mass components can be produced. This is attractive to all transport industries. However, titanium is inherently costly due to the high energy input required to extract this reactive metal from its ore. The combination of high cost and good specific properties are reflected in the greater use of titanium in higher end industries: there is more extensive usage of titanium alloys in military aerospace compared to civil aerospace.

Good corrosion resistance is another advantage of titanium alloys. Disadvantages are poor tribological performance and high cost. It must also be noted that the maximum operating temperature of titanium and its alloys is about 650 °C,

K. T. Voisey, *The Engineer's Guide to Materials*,
https://doi.org/10.1007/978-3-031-62937-2_3

with some variation from alloy to alloy. This is reasonably high and certainly allows Ti alloys to be used in some engine applications, but is notably lower than the 1668 °C melting point of Ti. This is due to two things: phase transformations and the fact that Ti can burn.

Titanium and its alloys are considered here. It is worth noting that titanium aluminide, TiAl, is a very different material. This does contain titanium and aluminium, however it is an intermetallic, a very different type of material to other titanium or aluminium alloys. TiAl is included in Chap. 11 Other Materials.

3.2 Sustainability Considerations Relating to Titanium and Titanium Alloys

The large energy input required to separate titanium from its ore is a clear concern when considering how sustainable use of titanium and its alloys is. However, the superior specific strengths that titanium alloys possess mean that there can be significant weight savings achieved by replacing steel or aluminium alloy components with titanium alloys. In the transport industry such weight saving is associated with decreased fuel consumption, which improves sustainability. In addition to this, the good corrosion resistance of titanium and its alloys lead to long service lives, again improving sustainability. The energy intensive titanium production process together with its good corrosion resistance mean that recycling of titanium is a cost effective alternative to using new titanium. The use of additive manufacturing processes can also impact the sustainability of titanium as the buy to fly ratios can be significantly improved, dramatically decreasing the amount of material wasted. Here both the saving in material usage and energy consumption by the additive manufacturing process have to be considered.

3.3 Applications of Titanium

The aerospace industry is the main consumer of titanium, using approximately 75% of it. Titanium's good specific properties make it very attractive. The higher material cost can be accommodated due to the low mass, and hence decreased associated fuel consumption, of the components produced. Another advantage of titanium alloys is that they have a fatigue limit. This is particularly relevant to applications such as rotating components that experience fluctuating stresses. However, as long as stress amplitudes are kept below the limit then fatigue will not occur. Fatigue is covered in more detail in Chap. 8.

Titanium alloys have both structural and engine applications in aerospace. All gas turbine engines flying today contain a significant amount of titanium alloys. In civil aviation the main use of titanium alloys is in the engines. Military aviation makes greater use of titanium alloys in structural applications. Titanium alloys, unlike aluminium alloys, do not experience galvanic corrosion issues when joined to carbon fibre reinforced polymer, CFRP.

Titanium alloys form 7% of the structure of the Boeing 777, 15% of the Boeing 787 [1], 6% of the Airbus A380 [2] and 14% of the Airbus A350 [3]. The usage is mainly in landing gear components [4]. Ti-10V-2Fe-3Al is used in various landing gear struts and links on the Boeing 777 (Fig. 3.1). The Boeing 787 uses the more advanced titanium alloy, Ti-5Al-5V-5Mo-3Cr for similar applications [5]. The most widely used titanium alloy in aerospace, Ti-6Al-4V, also has landing gear beam applications where it has been used for the Boeing 737, 747 and 757. Such applications exploit the good corrosion resistance of titanium alloys, required due to the exposure to the weather of these external components. The high specific strength is also relevant here as this means that the landing gear have smaller volumes, resulting in less loss of valuable payload space when they are retracted. There are additional advantages in the lower drag and noise levels generated by smaller landing gear assemblies. The Airbus A350 makes use of titanium alloys for engine pylons, landing gear and high load frames (Fig. 3.2) [6]. Ti-6Al-4V has been used within engine pylons in the Airbus A380, A330 and A350 [7]. On the Airbus A380 use of titanium alloy, instead of steel, pins within the pylon structure produced a 15 kg weight saving [2]. Airbus uses 3D printed titanium alloys brackets within the Airbus A350 XWB's pylon structure which attaches the engine to the wing.

The good specific properties of titanium alloys, combined with their good corrosion resistance, make them well suited to landing gear applications. They are a lower mass alternative to HSLA steel. The savings in fuel costs, as well as the decreased maintenance costs due to improved corrosion performance, make them an economically viable alternative.

Fig. 3.1 Boeing 777 landing gear, main strut is high strength steel, Ti-10V-2Fe-3Al is used in the supporting struts. "Boeing 777–200 starboard main gear leg, doors, bogie" by wbaiv is licensed under CC BY-SA 2.0

Fig. 3.2 An Airbus A350, this makes use of titanium alloys in engine pylons, landing gear and high load frames. "Airbus A350 XWB" by Joao Carlos Medau is licensed under CC BY 2.0

The different priorities of military aviation see wider use, with the material being used in both airframe structure as well as the engines. The alpha alloy, Ti-6Al-2Zr-2Sn-2Mo-2Cr-0.25-Si, was used in the airframe of the F22 and JSF (Joint Strike Fighter) projects/aircraft (Figs. 3.3 and 3.4). Ti-6Al-4V was used for both wings and body on the F22. The SR-71 Blackbird was well known for extensive use of titanium: over 90% of the structure was titanium. The Blackbird's flight speeds of Mach 2.5 induced aerodynamic heating to approximately 300 °C, exceeding the capacity of aluminium alloys. The specific titanium alloy used in the Blackbird was Ti-13V-11Cr-3Al, the combination of good corrosion resistance, high specific strength and good ductility made it a suitable choice. Military aircraft, including the F-22, C-17, F-35, and the UH-60 Black Hawk helicopter use a large volume of titanium alloy fasteners, as do remotely operated UAVs [8]. Ti-8V-6Cr-4Mo-4Zr-3Al is one titanium alloy used in aircraft fasteners, Ti-25V-3Cr-3Al-3Sn is another [4]. The improved corrosion performance, particularly when used with CFRP, compared to aluminium alloy or steel alternatives is an advantage here, though the weight saving is the main advantage [4].

Titanium fasteners are also used in civil aviation, particularly in areas where corrosion is a concern such as the toilets and galley areas. In 2022 it was reported that the global aerospace titanium fasteners market was valued at US$1,119M [8].

In order to minimise mass, gas turbine engines use titanium alloys as much as possible. Only switching over to the higher density nickel based superalloys when required by higher temperatures. This means that engine applications of titanium alloys are primarily in the front of the engine, where the operating temperatures are less severe. Engine applications of titanium alloys include the most highly visible components: the large fan blades that are seen at the front of the engine (Fig. 3.5). To decrease mass and centripetal forces, fan blades are hollow, made

Fig. 3.3 “F22 Raptor-RIAT 2017” by Airwolfhound is licensed under CC BY-SA 2.0

Fig. 3.4 “Lockheed Martin F-35B ‘Lightning II’ Joint Strike Fighter (JSF)” by aeroman3 is marked with Public Domain Mark 1.0

from a combination of specialist forming processes such as diffusion bonding and superplastic forming. Whilst the temperatures experienced here are not severe, the fan blades need to be strong. The rotation rates of up to 3300 rpm lead to centripetal forces of about 900 kN, almost 100 tonnes, on each blade.

Moving through the engine, titanium alloys are used for discs and compressor blades. These are rotating components where the high specific strengths of titanium alloys make them well suited to cope with the extreme stresses generated by centrifugal forces. Engine applications include the 3D-printed titanium alloy nozzles in the GE9X. However, once the operating temperature reaches about 600 °C

Fig. 3.5 The large Ti6Al4V fan blades of the Trent 1000 engine. **a** Fan blades visible at front of engine. **b** Close up of fan blade assembly, complex fan blade geometry is visible

titanium alloys are no longer suitable. The precise upper operating temperature varies from alloy to alloy.

Titanium has a combination of material properties which add up to it being an attractive spring material. Its modulus, strength and density mean that titanium springs can be 75% of the height and 60–70% of the weight of equivalent steel springs. The Volkswagen Lupo FSI and Ferrari's 360 Stradale both use Timetal LCB (a low cost beta alloy) for this application [9].

This makes them attractive for aerospace springs. There are also titanium springs used in bicycle and automotive applications. High performance bicycles use titanium frames (Fig. 3.6), here Ti3Al2.5V is often used as it has a good balance between strength required in service and the formability and weldability needed for manufacture. It also has appropriate corrosion resistance for bike frames and similar applications such as tennis rackets and golf club shafts [10].

Other automotive applications include exhaust systems, where the good corrosion resistance of titanium alloys, and again the mass reduction compared to stainless steel, make them attractive. That is as long as the temperature does not get too high, that is above approximately 400 °C [11]. These components do not need to carry high loads, and require good corrosion resistance, hence the lower strength, but cheaper and highly corrosion resistant commercially pure titanium, rather than an alloy, is used [12]. This use of the pure, rather than alloyed, material is unusual amongst metallic materials. Additional applications also using commercially pure titanium include chemical processing applications and heat exchangers. These applications exploit the superior corrosion resistance of the pure material

Fig. 3.6 Ti3Al2.5V is used in the frame of this high performance mountain bike. "Lynskey titanium singlespeed mountain bike with belt drive" by Richard Masoner/Cyclelicious is licensed under CC BY-SA 2.0

and again do not require the higher specific strengths of the titanium alloys. Commercially pure titanium has also been used in laptop housings where the high specific strength is sufficient to produce low mass but strong housings. This results in thinner and lighter laptops.

There are various marine applications of titanium alloys where strength as well as corrosion resistance is required. These include ship propellors, pressure shells for submarines and diving vessels [13], where the non-magnetic nature of titanium alloys can be advantageous in stealth terms, and deep sea petroleum production risers [14]. Titanium's good corrosion resistance, particularly in chloride containing conditions, is also exploited in the use of titanium in various chemical processing applications such as desalination plants [14], in heat exchangers in power plants, ships, offshore platforms and other areas in which the cooling medium is either seawater or impure water [15].

In 2020 the International Space Station upgraded its toilet facilities. The new system, the UWMS, the Universal Waste Management System, uses titanium as its good corrosion resistance makes it compatible with the acid used in urine pretreatment, as well as being able to deliver a low mass toilet [16].

Architectural applications of titanium famously include the Guggenheim Museum in Bilbao (Fig. 3.7) where titanium was selected as being able to produce a lower mass skin than competitor materials, as well as being strong and corrosion resistant enough to withstand continual exposure to weather.

Titanium also has land based military applications. There are various armour applications due to titanium's good ballistic characteristics, corrosion resistance and again its good strength to weight ratio which increases vehicle manoeuvrability. The M2 Bradley tank (Fig. 3.8), or more formally, infantry fighting vehicle, as currently used by the US features a titanium alloy hatch, which has a 35% mass saving compared to the previous aluminium alloy design [15].

Fig. 3.7 Ti cladding on the Guggenheim Museum, Bilbao. Photo courtesy of Liam Barnes

Fig. 3.8 The M2 Bradley infantry vehicle uses a titanium alloy hatch to save weight compared to the aluminium alloy alternative. "M2/A2 Bradley Fighting vehicles" by UNC-CFC-USFK is licensed under CC BY 2.0

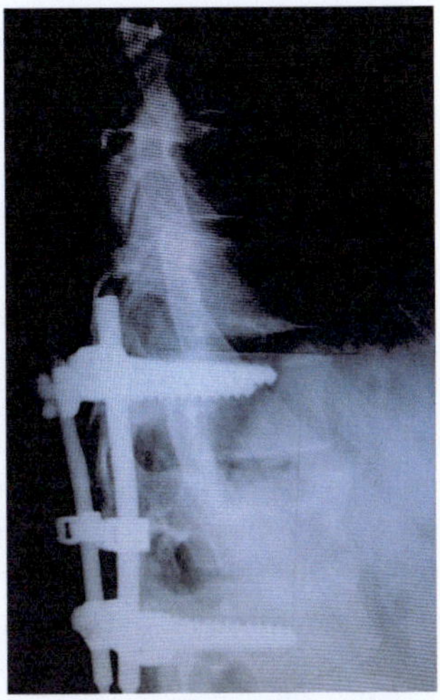

Fig. 3.9 Titanium rods and screws visible in an x-ray demonstrating some of the medical applications of titanium alloys. "Titanium pedicle screws" by warrenski is licensed under CC BY-SA 2.0

Its good biocompatibility, corrosion resistance and the good match in Young's modulus with bone make titanium a suitable material for medical use (Figs. 3.9 and 3.10). It is used in bone screws and anchors, dental implants and hip replacements as well as pacemaker cases. TI6Al4V is widely used in dental implants [17]. There are numerous applications in surgical devices including forceps, dental drills, endoscopes and scalpels. The non-magnetic nature of titanium and its alloys is advantageous for medical application as this makes them MRI, magnetic resonant imaging, compatible.

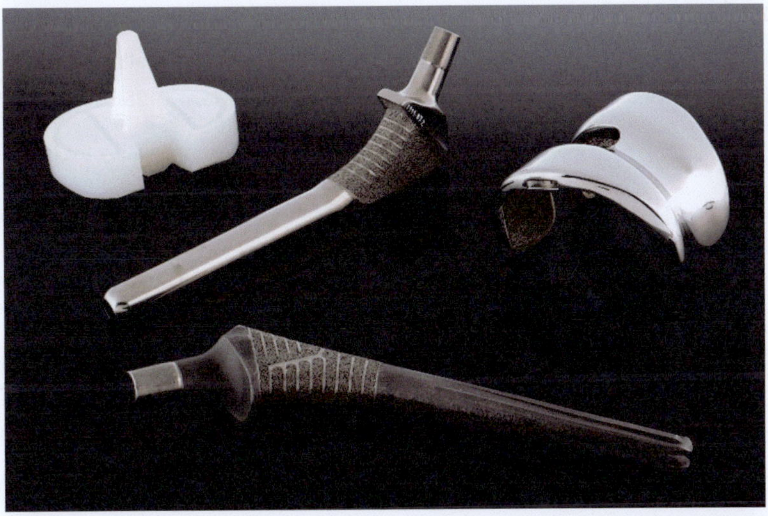

Fig. 3.10 The long metallic items are titanium alloy replacement hips. "File:Hip joint replacement, United States, 1998 Wellcome L0060175.jpg" is licensed under CC BY 4.0

3.4 What Engineers Need to Know About Titanium

Titanium has a good set of properties and characteristics. The relatively low density mean that it can have great specific properties, making it very attractive to the aerospace industry. The low modulus also means that it has spring applications. However, all is not quite as good as it first appears. The very attractive high melting point of 1668 °C cannot be properly exploited. Phase transformations and the risk of burning limit the maximum operating temperature to about 600 °C, with some variation depending on which particular alloy is considered. Titanium is another inherently expensive material. This, like aluminium, is because it is highly reactive and hence requires a lot of energy to separate the metal from the ore. Again, as with aluminium, this reactivity is useful in generating a largely protective surface oxide which gives good corrosion resistance.

Titanium can exist in two different phases, alpha and beta. These two phases mean that titanium alloys fall into one of three groups depending on what phases are present: alpha or beta or alpha and beta. The overall properties of the alloy depend on which phases are present (Table 3.1). For the two phase alpha and beta alloys there is the added complication of the properties depending on how the two phases are arranged. As is seen in many metallic materials, as soon as multiple phases are present there is scope to control material properties by manipulating how the different phases are arranged. This can be done by heat treatment. As with aluminium alloys, significant changes in material properties can be achieved via heat treatment, without the material melting. Knowledge of the alloy's previous heat treatment history, in addition to the composition, is needed to understand what the material properties and performance will be.

Table 3.1 Summary of titanium alloy types

Alloy type	Characteristic properties	Example alloy and application
Alpha (α), and near alpha	Good creep resistance, not heat treatable, weldable, low ductility, maximum operating temperature 350–760 °C	Ti6Al2Zr2Sn2Mo2Cr0.25Si (6.22.22) Airframe alloy for F22 and JSF projects
Alpha-beta (α, β)	High strength, good formability, maximum operating temperature 315–480 °C	Ti6Al4V many applications in engines and airframes
Beta (β), and near beta	High formability, maximum operating temperature 288–816 °C	Ti10V2Fe3Al aircraft landing gear

3.4.1 How Titanium Alloys Work—The Impact of the Alpha and Beta Phases

Deliberate manipulation of the two different phases of titanium is central to understanding and effectively exploiting titanium alloys.

Pure titanium can exist in two forms. At temperatures below 883 °C the alpha phase is stable, whereas at higher temperatures it is the beta phase that is stable. This matters because the alpha and beta phases have different properties due to their different crystal structures. The alpha phase is stronger and has good creep resistance whereas the beta phase has a higher formability and is tougher due to the inherently higher ductility. It is almost always undesirable for phase changes to occur in service as this changes the material properties. Repeated cycling between different phases will lead to cracking and ultimately component failure.

In pure titanium the change between alpha and beta phases occurs at 883 °C: the material will simply be in one phase or the other depending on the temperature. For alloys the situation is more complicated. Not only does the phase transformation temperature vary with composition, but it is possible for both phases to be present at once. This is key to being able to deliberately manipulate and control the overall properties and hence performance of any titanium alloys component. The actual properties of any titanium alloy are a combination of the phases present and the microstructure. These in turn are a function of composition, heat treatment and mechanical processing.

Each of the two phases have different inherent properties due to their different underlying crystal structures. The alpha phase is the lower temperature form of titanium. It has a higher strength than the more ductile beta phase that is the high temperature form. Whilst it is often seen that yield strength decreases and ductility increases as temperature increases, here the beta phase has an inherently higher ductility than the alpha phase. The beta phase has a body centre cubic structure whereas the alpha phase has a hexagonally close packed structure. The body centre cubic structure is more ductile because it has a larger number of slip systems, meaning that there are more ways in which dislocations can move, making dislocation motion easier, which is equivalent to saying that the material is more ductile. This gives a step change increase in ductility when the alpha phase transforms into the beta phase. This difference in ductility can be exploited by carrying out forming operations in the beta phase. Yes, a higher temperature is needed, but lower forces are required, hence the overall operation can be cheaper.

Alloying additions in titanium alloys fall into two main groups: alpha stabilisers and beta stabilisers. Alpha stabilisers, such as aluminium, stabilise the low temperature alpha phase to higher temperatures, allowing exploitation of the higher strength and superior creep resistance of the alpha phase. Beta stabilisers, including vanadium and molybdenum, stabilise the beta phase to lower temperatures, allowing wider use of the more ductile beta phase. Neither phase should be regarded as inherently superior to the other, just different, with the desirability of the properties varying depending on the particular application.

The most widely used titanium alloy is Ti6Al4V, often referred to simply at Ti64 (that is Ti six four, not sixty four). This has both an alpha stabiliser, Al, and a beta stabiliser, V, and is a two phase, alpha and beta alloy. The resulting balance of properties, and the ability to tailor the properties by manipulation of the two phase structure, has made Ti6Al4V a popular alloy. This is a relatively cheap titanium alloy which features good creep resistance, fatigue resistance, castability and strength. It can be used up to around 350 °C. However, care needs to be taken during any welding operations. Welding can be regarded as an extreme heat treatment. The thermal cycles induced in the weld and near weld material can override any previous heat treatment operations carried out deliberately to control microstructure and hence properties.

Ti6Al2Sn4Zr2Mo, or Ti6242 (Ti six two four two), has a higher maximum operating temperature of about 540 °C. This reflects the larger proportion of the alpha phase. The alloy is termed a "near alpha" alloy, indicating that alpha is the majority phase. The resulting increased creep resistance is exploited in gas turbine engines. Ti6Al4V has a lower maximum operating temperature of 350 °C so is generally restricted to fan blades and low pressure compressor components within the engine. However using Ti6242 means that parts of the high pressure compressor can also be made of Ti alloy, representing a valuable weight saving over alternative materials such as Ni superalloys. Here, the beneficial reduced fuel burn associated with the weight saving more than compensates for the inherent cost of the material. Other characteristics of alpha and near alpha alloys is their good creep resistance and the fact that they are weldable.

Phase stabilisation is not the only reason to alloy titanium. Solid solution strengthening can be achieved by alloying with elements including aluminium, tin and zirconium. Some elements have multiple roles, alloying with aluminium can also be done to further push down the density of the alloy. Chromium is also an important alloying element in titanium alloys, as well as acting as a beta stabiliser it improves fire resistance, hence increasing maximum operating temperatures and opening up a wider range of applications, particularly in engines.

3.4.1.1 Using Mechanical Processing with Heat Treatment to Manipulate Alpha Beta Alloys

The composition of the titanium alloys determines which phase or phases are present. For two phase alloys, such as Ti6Al4V, heat treatment combined with mechanical processing can be used to create an array of different microstructures. The microstructure is the internal structure of the material, it describes how the different phases are arranged: the shape and size of the regions of each phase. The microstructure has a huge impact on material properties.

For Ti6Al4V 980 °C is an important temperature, at higher temperatures, above 980 °C only the beta phase is present. For applications where a large amount of deformation is required during manufacture of components it can be advantageous to carry out forming operations in the beta phase (Fig. 3.11). This exploits the higher ductility of the beta phase, meaning the deformation processes can be carried out under lower loads, and hence with cheaper equipment, though there will

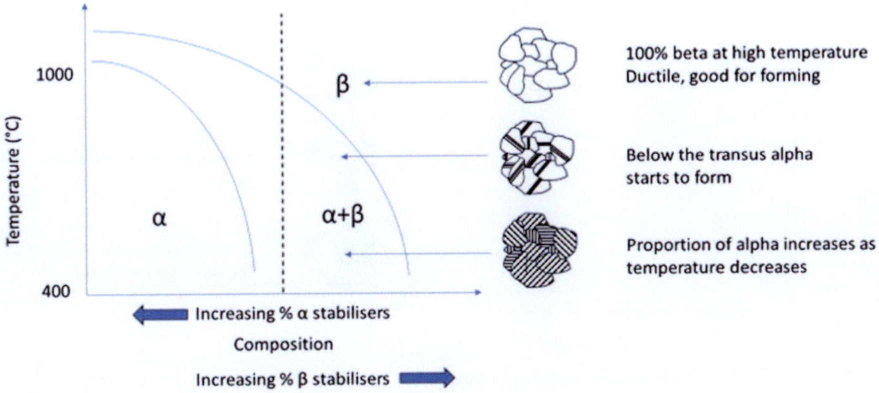

Fig. 3.11 Schematic phase diagram for titanium alloys, and corresponding schematic microstructures highlighting forming in the beta phase region, after [18]

be some cost related to the elevated temperatures required. After all required shaping operations have been carried out the material cools. As the temperature drops below 980 °C the alpha phase starts to appear. The alloy gets stronger as the two phase alpha–beta structure forms. A distinctive lamellar structure is formed, with alternating layers of alpha and beta. It is the interfaces generated between the different phases within the material which act as obstacles to dislocation motion, increasing the yield strength. The speed of cooling influences the arrangement of the phases. Faster cooling produces a finer structure, this has more interfaces per unit area and hence is a more effective obstacle to dislocation motion and a stronger material.

Alternatively, forming operations could be done within the two phase region (Fig. 3.12). In this case the material temperature would be raised to increase ductility, but not so high that the single phase, beta only, regime was reached. The material would be a mixture of alpha and beta regions during forming. After forming, as the material cools within the two phase alpha beta region the relative proportions of the two phases change, with the amount of alpha increasing. When the alloy has reached room temperature some of the beta regions will have transformed into the lamellar, alpha plus beta. The overall microstructure features regions with the lamellar structure interspersed with regions of the beta phase only. The higher ductility of the beta regions mean that the overall alloy has a higher ductility and fatigue strength but a lower fracture toughness and creep strength compared to the fully lamellar microstructure which can be obtained by different thermo-mechanical treatment of the same, Ti6Al4V, alloy.

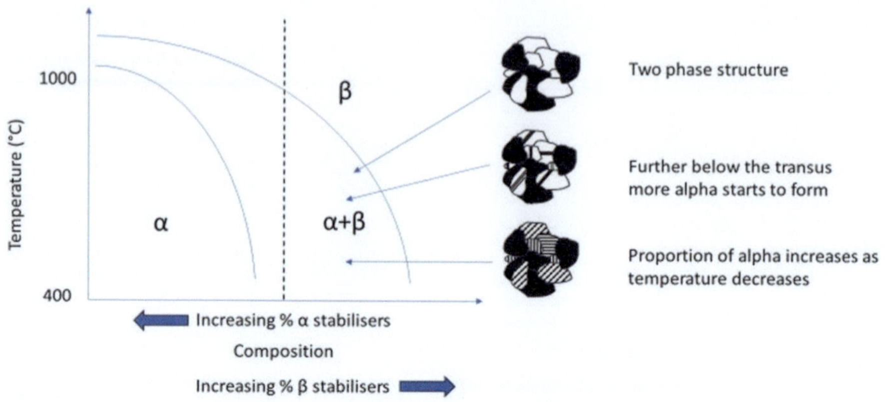

Fig. 3.12 Schematic phase diagram for titanium alloys, and corresponding schematic microstructures, highlighting forming in the two phase region, after [18]

3.5 Test Your Understanding—Questions

Q1 Ti alloys and Al alloys are both regarded as low density metallic alloys, but how do their densities compare to eachother?

Select one:

a. Al alloys are more dense
b. they are pretty much the same, no significant difference
c. Ti alloys are more dense

Q2 Titanium has a high melting point, Tm, of 1668 °C however Ti alloys cannot fully exploit this and their maximum operating temperature is a lot lower—why?

Select one:

a. Ti alloys melt very quickly so you need to ensure you stay a long way away from the melting temperature in service
b. Ti alloy usage at high T is limited by phase transformations and Ti burning, both happen at T < Tm
c. Ti alloys have a massive coefficient of thermal expansion and expand so much at high T that it is not practical to use them at these temperatures

Q3 Which industrial sector are the vast majority of titanium applications in?

a. Automotive
b. Aerospace
c. Marine
d. Manufacturing

Q4 Compare the material properties of a typical α titanium alloy and a typical β titanium alloy and state why they are different.

Q5 Ti6Al4V and Ti6Al2Sn4Zr2Mo are alpha–beta and near alpha titanium alloys respectively. Qualitatively, state and explain the differences in material properties and maximum operating temperatures.

Q6 What is the main factor restricting widespread use of titanium alloys in the automotive industry?

Q7 Qualitatively compare the properties of the two different versions of Ti6Al4V described below and explain the differences in prior thermal history that have produced these different microstructures.

- Ti6Al4V with a fine alpha/beta lamellar structure
- Ti6Al4V with a coarse alpha/beta lamellar structure

3.6 Test Your Understanding—Answers

Q1 Answer c Ti alloys are more dense.

Q2 Answer b Ti alloy usage at high T is limited by phase transformations and Ti burning, both happen at $T < Tm$.

Q3 Answer Which industrial sector are the vast majority of titanium applications in?

b. Aerospace

Q4 Answer Beta alloy would be more ductile, alpha alloy would be higher strength, difference due to the different crystal structures.

Q5 Ti6Al4V is an alpha–beta-2 phase alloy so properties can be controlled by heat treatment, however this can lead to problems with welding. The alloy will have a lower alpha/beta transus on the phase diagram due to the presence of beta stabilising V, and hence has a lower maximum operating temperature than Ti6Al2Sn4Zr2Mo. Ti6Al2Sn4Zr2Mo is a near alpha alloy so has a high proportion of the alpha phase, this is less ductile than the beta phase hence this alloy will have better creep resistance and lower ductility than the Ti64, it is also not heat treatable so can easily be welded.

Q6 Answer Cost, this is a particular issue for the very price sensitive automotive industry.

Q7 Answer The fine structure version of Ti6Al4V will have a higher yield strength and lower ductility, this is due to the greater number of interfaces, and hence greater number of obstacles to dislocation motion, per unit area. The finer structure will have been produced by more rapid cooling from the beta phase region.

Further Reading

Organisations and companies

The International Titanium Association is a membership based trade organisation. Various titanium related resources are available through their website, more are available after free registration is completed. https://titanium.org/
TIMET is a leading titanium producer, their website has information on different applications as well as their global operations https://www.timet.com/
Nippon Steel have produced a technical report on titanium in the aerospace industry. Nippon Steel & Sumitomo Metal Technical Report No. 106 JULY 2014 https://www.nipponsteel.com/en/tech/report/nssmc/pdf/106-05.pdf

Open access articles and research papers

Extending the Use of Titanium Alloys on A350XWB, S. Audion, G. Khelifatii, J. Delfosse, P. -Y. Oillic, R. Peraldi Proceedings of the 12th World Conference on Titanium https://cdn.ymaws.com/titanium.org/resource/resmgr/ZZ_WTCP_2011_Re-Do/V3/2011_Vol.3-3-Extending_the_U.pdf good overview of usage in Aibus/aero applications.
Opportunities and Issues in the Application of Titanium Alloys for Aerospace Components by James C. Williams and Rodney R. Boyer *Metals* **2020**, *10*(6), 705; https://doi.org/10.3390/met10060705.
Titanium and Civil Aviation: Turbulence Ahead, By Stainless Steel World Publisher—May 11, 2022 https://stainless-steel-world.net/titanium-and-civil-aviation-turbulence-ahead/

https://www.sciencedirect.com/science/article/pii/S2452321619301738 open access paper on “Effect of Temperature and Cooling Rates on the $\alpha + \beta$ Morphology of Ti-6Al-4V Alloy”.

References

1. Forty Years of Structural Durability and Damage Tolerance at Boeing Commercial Airplanes Steven A. Chisholm, Antonio C. Rufin, Brandon D. Chapman and Quentin J. Benson, Boeing Technical Journal, Innovation Quarterly, May 2017, vol 1 Issue 4 https://www.boeing.com/resources/boeingdotcom/features/innovation-quarterly/may2017/btj_strutures_full.pdf

2. Extending the Use of Titanium Alloys on A350XWB S. Audion, G. Khelifatii.,]. Delfosse, P. -Y. Oillic, R. Peraldi Proceedings of the 12th World Conference on Titanium https://cdn.ymaws.com/titanium.org/resource/resmgr/ZZ_WTCP_2011_Re-Do/V3/2011_Vol.3-3-Extending_the_U.pdf
3. Aircraft and Engine Suppliers are Replacing Russian Titanium, Sean Broderick, August 15 2022, Aviation week
4. New titanium applications on the Boeing 777 airplane, Rodney R. Boyer M.S., JOM 44, 23–25 (1992). https://doi.org/10.1007/BF03223045
5. Attributes, Characteristics, and Applications of Titanium and Its Alloys, R.R. Boyer, JOM May 2010 Vol. 62, No.5 pp. 35–43 https://www.tms.org/pubs/journals/jom/1005/boyer-1005.html
6. https://flywith.virginatlantic.com/eu/en/stories/the-airbus-a350-a-quiet-efficient-giant.html#:~:text=Aerodynamics,different%20phases%20of%20the%20journey accessed 22/10/23
7. Substituting Ti-64 with AA2099 as Material of a Commercial Aircraft Pylon, Hamza Khalid, A.A. Gomez-Gallegos, Advances in Materials Science, Vol. 21, No. 2(68), June 2021, https://doi.org/10.2478/adms-2021-0012 which reported information from the following four sources: Structural blind fasteners. Flight Airworth Support Technology, (27) (2019) 26–9; Ferrer G, Chamfroy C, Dupouy. SS. A350 XWB composite repairs. Flight Airworth Support Technology, (2016) 32. 28; Pora J. Advanced materials and technologies for A380 structure. Flight Airworth Support Technology, (2003) 32. 29.; A320 family (CFM56) familiarization course. Islamabad, Pakistan (2018).
8. Insights on the Aerospace Titanium Fasteners Global Market to 2028 - Increasing Demand of Light Weight Fasteners Presents Opportunities - ResearchAndMarkets.com June 21, 2022 10:26 AM Eastern Daylight Time https://www.businesswire.com/news/home/20220621005868/en/Insights-on-the-Aerospace-Titanium-Fasteners-Global-Market-to-2028---Increasing-Demand-of-Light-Weight-Fasteners-Presents-Opportunities---ResearchAndMarkets.com accessed 28/11/23
9. Froes, F.H., Friedrich, H., Kiese, J. et al. Titanium in the family automobile: The cost challenge. JOM 56, 40–44 (2004). https://doi.org/10.1007/s11837-004-0144-0
10. https://www.azom.com/article.aspx?ArticleID=1728 accessed 28/11/23
11. Titanium for Automotive Applications, Jun 20 2001, updated 4th Feb 2020 by Clare Kennan, AZoM https://www.azom.com/article.aspx?ArticleID=553
12. Applications of Titanium alloy in the Automobile Industry, RefractoryMetal.org accessed 18/11/23 https://www.refractorymetal.org/applications-of-titanium-alloy-in-automobile-industry/
13. Titanium in Marine Application, StandardTitanium accessed 18/11/23 https://titanium.net/titanium-in-marine-application/#:~:text=Deep%20diving%20vehicles%20and%20submarines,shells%20for%20deep%2Dsea%20equipment.
14. Titanium – Applications March 4 2002 AZoM https://www.azom.com/article.aspx?ArticleID=1297#:~:text=Titanium%20alloys%20are%20used%20in,heat%20exchangers%2C%20and%20hydrometallurgial%20autoclavesaccessed 28/11/23
15. https://titanium.com/markets/industrial/#:~:text=Titanium%20provides%20an%20economically%20efficient,for%20Chemical%20Processing%20Industry%20applications. Accessed 28/11/23
16. Nature calls: Astronauts take $23 million titanium toilet to International Space Station, Sarwat Nasir, Oct 04 2020 https://www.thenationalnews.com/uae/science/nature-calls-astronauts-take-23-million-titanium-toilet-to-international-space-station-1.1087436 accessed 18/11/23
17. Liu X, Chen S, Tsoi JKH, Matinlinna JP. Binary titanium alloys as dental implant materials-a review. Regen Biomater. 2017 Oct;4(5):315–323. https://doi.org/10.1093/rb/rbx027. Epub 2017 Sep 23. PMID: 29026646; PMCID: PMC5633690.
18. D is for Duplex Titanium Alloys, Part 2, Dr Mark J Whiting, Nov 20 2016, MetalsAndAlloysBlog https://metalsandalloysblog.wordpress.com/tag/phase-diagrams/ accessed 15/12/2023

Superalloys 4

Superalloys the bullet points

- Superalloys are high strength metal alloys which retain their strength to high temperature, making them well suited to applications within the most demanding parts of gas turbine engines
- Nickel based superalloys are the main type of superalloys due to their combination of properties, iron and cobalt superalloys also exist
- The superior properties of nickel based superalloys are due to a particularly effective form of precipitate strengthening, the γ/γ', gamma/gamma prime structure
- Superalloys have a large number of alloying elements, typically about ten

4.1 Introduction

Superalloys were designed to be used in gas turbine engines. They are used for components which operate in the most demanding conditions of a high temperature, high pressure, corrosive environment with rotating components subject to extreme stresses due to the high rates of revolution. The superior properties of superalloys combine to allow them to operate in these demanding operating conditions. A disadvantage is their relatively high density of approximately 8900 kg m^{-3}, higher than steel (7870 kg m^{-3}) and almost double that of titanium alloys (4500 kg m^{-3}). However, the other properties of nickel based superalloys, particularly their high strength at high temperature and good creep resistance, more than compensate for this. The usefulness of superalloys comes from the drive for higher engine operating temperatures and the associated improved efficiencies.

Superalloys can be iron, cobalt or nickel based. Nickel based superalloys dominate due to their overall combination of properties.

K. T. Voisey, *The Engineer's Guide to Materials*,
https://doi.org/10.1007/978-3-031-62937-2_4

It is often reported that superalloys operate above their melting points. That is not true. What is true is that they operate in environments where the air temperature exceeds their melting points. A lot of engineering effort in the form of insulating thermal barrier coatings and active cooling ensures that the metal temperature remains below the melting point. This enables nickel based superalloy components to survive in a hot gas path that may reach 2000 °C, significantly higher than the 1455 °C melting point of nickel.

Superalloys have a large number of alloying elements, typically ten. This can include some rare earth elements which can lead to issues about security of material supply.

There is a drive to replace superalloys with alternative materials which allow even higher operating temperatures to be achieved. This is an area of active research, with real advances being made, such as the introduction of intermetallic and ceramic matrix composite components into gas turbine engines.

4.2 Sustainability Considerations Relating to Superalloys

The production of nickel itself is energy intensive. The energy input is further increased by the alloying, specialised casting facilities and heat treatments needed in the manufacture of nickel based superalloy components. However, the Nickel Institute states that the overall impact of all uses of nickel is to decrease greenhouse gases [1].

Nickel based superalloys feature a large number of alloying elements, including rare earth elements such as yttrium. There are sustainability concerns related to these elements due to the fact that they are rare, the processes used to extract them are complex and produce large amounts of toxic waste. Other alloying elements such as hafnium, cobalt, tantalum and tungsten are included on the European Union's list of critical raw materials, meaning that they are materials of significant economic importance to the EU and which are at high risk of having supply issues [2]. In addition to this, rhenium and ruthenium which enhance creep strength in modern superalloys, are both rare metals. This means that there is a lot of interest in recycling of superalloy scrap, which, albeit driven by economic reasons, does improve the sustainability of the material.

The main application of superalloys is in gas turbine engines. Their use and development has allowed engine operating temperatures to be pushed higher, increasing engine efficiency. This improves the overall sustainability of operating aircraft as fuel burn is decreased.

4.3 Applications of Superalloys

Gas turbine engines are by far the main application of superalloys. Typically 40–50% of the weight of an aeroengine is attributable to nickel based superalloys. Nickel based superalloys are not visible in Figs. 4.1 and 4.2 because they are used

Fig. 4.1 A GEnx engine. "Vietnam Airlines 787-10 VN-A879 GEnx-1B nacelle" by Dylan T is licensed under CC BY 2.0. To view a copy of this license, visit https://creativecommons.org/licenses/by/2.0/?ref=openverse

Fig. 4.2 A RR Trent 100 engine

in the most demanding areas, inside the engine along the hot gas path, where the temperatures are highest.

The blades visible in Fig. 4.2 are the large fan blades, these are made from Ti6Al4V and are approximately 1 m in length. The nickel based superalloy turbine blades within the engine are much smaller, approximately 10 cm in length (Fig. 4.3).

The rear part of gas turbine engines are dominated by nickel based superalloys (Fig. 4.4). The conditions are particularly demanding, not only are there high temperatures and pressures but also an aggressive, corrosive atmosphere due to the combustion gases. In addition to this there are a lot of rotating components. Compressor discs, compressor blades, turbine discs and turbine blades are all rotating at thousands of revolutions per minute. The consequent centripetal forces mean that these components are highly stressed and require a material that can withstand extended loading at high temperatures.

Nickel based superalloys are frequently used in the hotter, higher pressure, end of the compressor, where the temperatures become too high for titanium alloys to

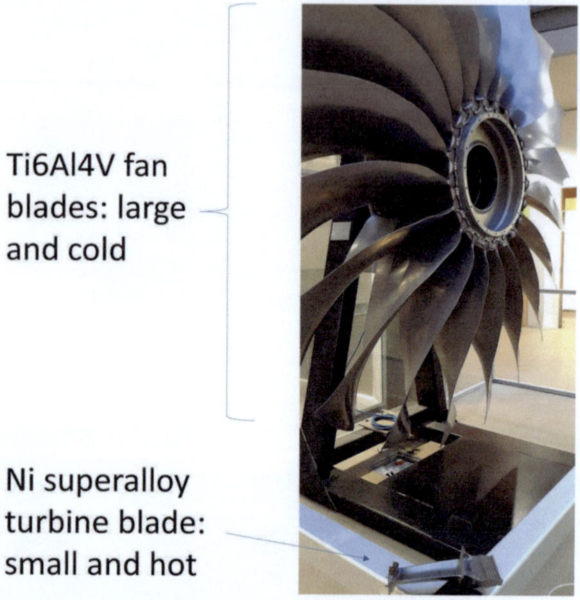

Fig. 4.3 Despite having a similar name and shape, fan blades and turbine blades have very different dimensions and operating conditions. The fan blades here are Ti6Al4V, chosen because of its high specific strength. The turbine blade is made from a nickel based superalloy, this is used as titanium alloys cannot cope with the elevated temperatures encountered within the turbine

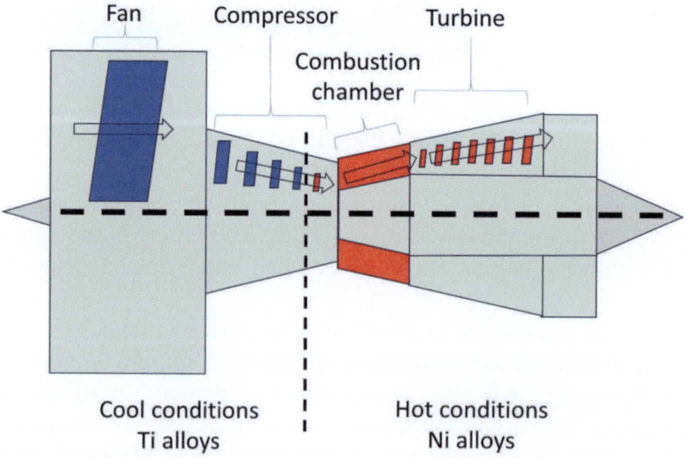

Fig. 4.4 Schematic illustration of a gas turbine aeroengine. Arrows indicate gas path through engine. The front of the engine is relatively cool and dominated by titanium alloys, the vertical dotted line indicates the approximate position at which the temperature is too high for titanium alloys and there is then a switch over to nickel based superalloys for the rear, hot, part of the engine

be used. Inconel 718 has been used for compressor and turbine discs. Nickel based superalloys are routinely used for combustion chambers, turbine blades, nozzle guide vanes and turbine discs. The GEnx engine, used on the Boeing 787 has, like most engines, a nickel superalloy combustion chamber and turbine blades. High pressure turbine discs use nickel based superalloys, due to their strength at temperature and creep resistance. U720Li and RR1000 are disc alloys used in the Trent 900 and Trent 1000 engines. RR1000 is used in the high pressure turbine disc. CMSX4 is a single crystal superalloy developed for use in aeroengine turbine blades.

There are also some applications in automotive engines such as in turbo chargers where their superior corrosion resistance is exploited.

The superior corrosion resistance and high temperature performance of nickel based superalloys also leads to applications in the petrochemical refining industry. Inconel 625 has been successfully used in refinery distillation towers, including nozzle applications where the high temperature acidic conditions require exceptional corrosion resistance [3]. Inconel 625 has also been used in waste incineration systems, and marine exhaust ducts, where the corrosion resistance, particularly the resistance to chloride containing environments at temperature is exploited [3, 4]. The corrosion resistance of Inconel 625 in marine environments, as well as its high tensile strength has led to use of this alloy in marine mooring cables and propeller blades [4].

4.3.1 What are Superalloys?

Superalloys are metallic alloys. There are three types: iron, cobalt and nickel based superalloys. Of these, nickel based superalloys are by far the most prevalent due to their combination of properties. There are many nickel based superalloys, the names of which, CMSX4, Inconel 718, Nimonic 80A, RR1000... show which "family" the alloy belongs to and which typically reflect the original manufacturer: Cannon Muskegon and International Nickel Company respectively for CMSX4 and Inconel 718, RR stands for Rolls-Royce plc (Table 4.1).

Having a large number of alloying elements, typically about ten, is a typical characteristic of nickel based superalloys. Some are present to enhance a specific property, some have multiple roles. What follows is an illustrative but non-exhaustive list.

Chromium, aluminium, hafnium, silicon and yttrium are all used to enhance corrosion resistance. Chromium, aluminium and silicon can enhance the exterior oxide layer whereas adhesion of that layer is improved by yttrium and hafnium. Tungsten, molybdenum, rhenium, iron, aluminium, chromium, cobalt, niobium, tantalum, vanadium and titanium can all provide solid solution strengthening. Grain boundary strengthening can be enhanced by boron and zirconium. Magnesium, silicon, carbon, chromium, molybdenum, tantalum, hafnium, tungsten, niobium and titanium may all contribute to carbide strengthening. Iron may be

Table 4.1 Compositions of example nickel based superalloys

Weight %	Nimonic 80a	CMSX4Single crystal	Inconel 625	RR1000 [5]
Ni, Nickel	69	61.7	58	Bal
Cr, Chromium	18–21	6.5	20–23	15
Ti, Titianium	1.8–2.7	6.5	0.40 max	3.6
Al, Aluminium	1–1.8	1	0.40 max	3
Fe, Iron	< 3		5.0 max	
Co, Cobalt	< 2	6.5	1.0 max	18.5
Si, Silicon	< 1		0.5 max	
Zr, Zirconium	< 0.15			0.06
C, Carbon	< 0.1		0.1 max	0.03
S, Sulphur	< 0.015		0.015 max	
B, Boron	< 0.008			0.03
W, Tungsten		6.4		
Re, Rhenium		3		
Ta, Tantalum		6.5		2
Hf, Hafnium		0.1		0.5
Mo, Molybdenium			8–10	5
Nb(+Ta), Niobium			3.15–4.15	
Mn, Manganese			0.5 max	
P, Phosphorous			0.015 max	

included simply to decrease overall cost. By far the most important alloying element is aluminium as this is key to the formation of the gamma/gamma prime structure responsible for the impressive high temperature strength. Titanium, tantalum and niobium can also contribute to this. Rhenium has been a feature of more recent generations of superalloys, where it enhances creep strength.

Single crystal superalloys have no need for grain boundary strengtheners and omission of these elements leads to an advantageous increase in melting point and larger extent of gamma prime, with associated improvement in material properties. Further development of single crystal nickel based superalloy composition with time has seen addition of Tungsten and Rhenium to improve solid solution strengthening and creep resistance. Later generations have seen further increase in first Rhenium and then Ruthenium content to further increase creep resistance.

As each alloying element enhances at least one property it is logical to assume that the more alloying elements the better. However, that is not the case. Alloy designers add as many alloying elements as they can before properties no longer improve. Topologically close packed, TCP, phases can form, and be detrimental to overall alloy properties and performance, if alloying additions are present in excessive quantities. TCP phases are problematic as they are brittle, platelike, phases which are undesirable as they act as stress concentrators. They also deplete the main matrix of otherwise useful alloying additions. Here it is important to note that casting operations may result in non-uniform elemental distributions, hence care must be taken with heat treatments to allow compositional uniformity.

4.3.1.1 How Superalloys Work—The γ/γ' Structure

The gamma/gamma prime structure underlies the impressive behaviour of nickel based superalloys and gives them their very distinctive, and easy to sketch, microstructure of cuboidal precipitates (Figs. 4.5 and 4.6). This is basically precipitate strengthening at its best. Gamma prime refers to the precipitates, with gamma being the nickel in between.

The gamma prime precipitates are particularly effective at increasing strength due to three reasons:

- They are coherent precipitates, which gives good high temperature stability
- They have an ordered structure, which makes them an inherently better obstacle to dislocations

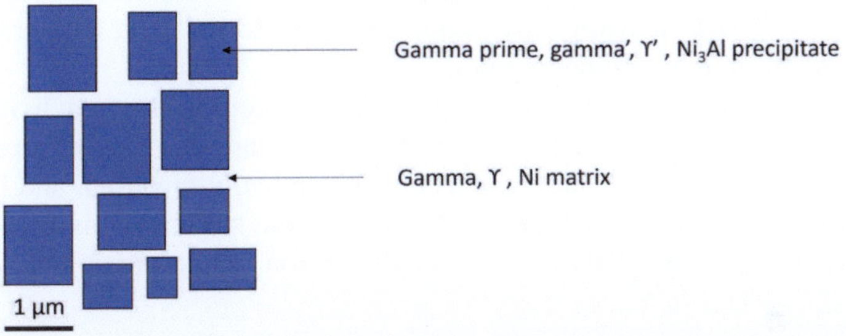

Fig. 4.5 Schematic illustration of the gamma/gamma′, γ/γ', structure typical of nickel based superalloys

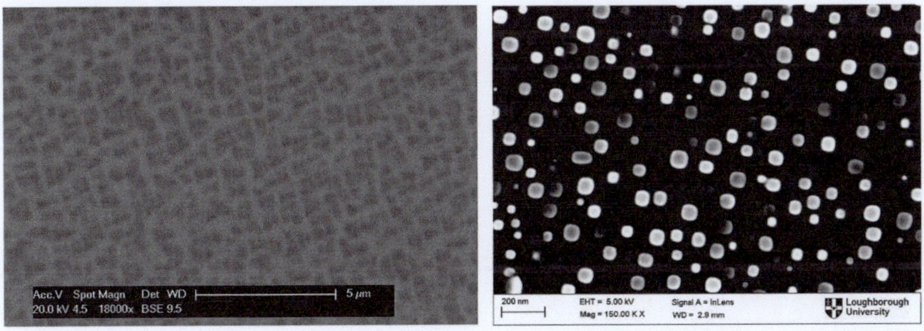

a b

Fig. 4.6 Scanning electron micrographs showing the gamma/gamma′ structure typical of nickel based superalloys. **a** C1023, courtesy of Cletus Akisin. **b** is courtesy of Simon Hogg, Loughborough University

- The yield strength, or yield stress, anomaly

Coherency means that there is good match in crystal structure between the precipitates and the surrounding matrix (Fig. 4.7a). This means the two crystal structures fit together well and the interface has a low associated energy. This results in the cuboidal shapes of the gamma prime, γ', precipitates. The planes in the two crystal structures which match well form the surfaces of the precipitates. Coherency is beneficial as it makes the precipitates stable against thermally induced coarsening. Incoherent precipitates have a poor match between the crystal structures of the matrix and precipitates (Fig. 4.7b). The resulting high energy interfaces between the precipitates and matrix mean there is a large driving force for coarsening (Fig. 4.7d). Coarsening decreases the total interfacial area, this is thermodynamically desirable as it decreases the total energy of the system. Coarsening is thermally activated as it relies on diffusion. It needs to be minimised for any material that has to operate at high temperature. With gamma prime and gamma having coherent interfaces this means that coarsening is not a problem, i.e. once the gamma/gamma prime structure is generated it is retained (Fig. 4.7c). Without coherent interfaces the gamma/gamma prime structure and associated beneficial properties are at risk of being lost during exposure to the high temperatures that are encountered in gas turbine engines.

The ordered structure of the gamma prime precipitates is also beneficial. Having an ordered structure means that the Ni and Al atoms are in a regular arrangement. This arrangement forms because it is thermodynamically favoured, and hence low energy. However, if a dislocation moves through the structure this favoured, low energy structure, is lost. This means that the region of the precipitate which has

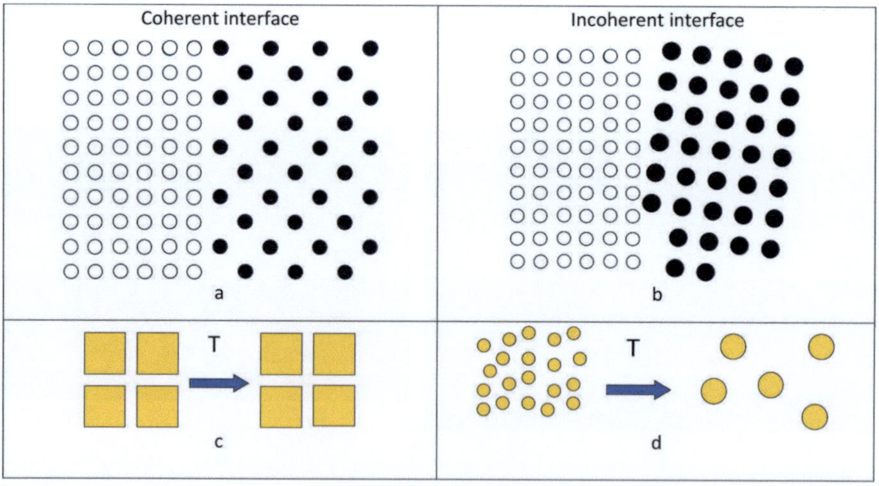

Fig. 4.7 Schematic diagrams showing the difference between coherent and incoherent interfaces and the effect on precipitate distribution stability on exposure to high temperature

had a dislocation move past it, is now in a higher energy state (Fig. 4.8, top). For disordered structures, these are disordered both before and after the passage of a dislocation. There is no difference between energy levels for material where dislocations have or haven't passed (Fig. 4.8, bottom).

For the ordered material, if a second dislocation moves through the material this essentially fixes the material, returning it to the original, low energy structure (Fig. 4.9). There are then three areas of material: the original low energy material where no dislocations have passed, the "repaired" low energy material where two dislocations have passed, and between them the high energy region where a single dislocation has passed (Fig. 4.10).

This now means that there is a thermodynamic driving force for the two dislocations to be as close together as possible as this would lower the energy of the system. However, as the separation between the dislocations decreases the stress/strain fields around each dislocation start to overlap (see Chap. 12). As the overlap increases the total stored strain energy in the system increases as this scales with the square of the strain. The dislocations end up at an equilibrium separation where the energy is minimised. The two dislocations now move together in order

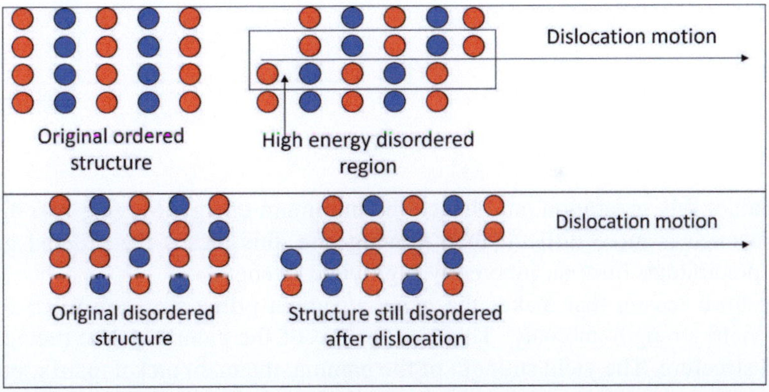

Fig. 4.8 Effect of dislocation passing through (top) an ordered structure and generating a high energy, disordered, region, (bottom) a disordered structure to produce a changed but still disordered structure

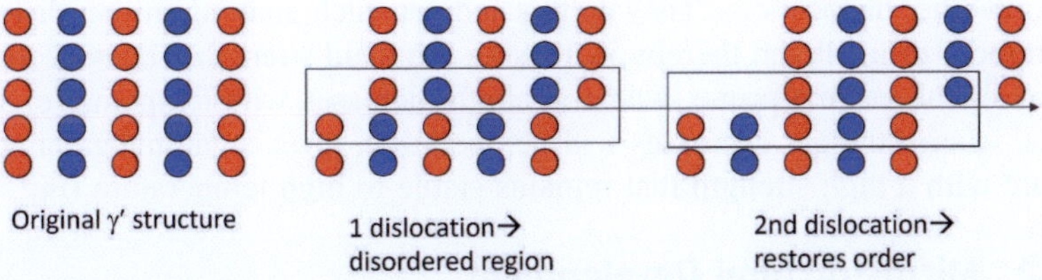

Fig. 4.9 Effect of two dislocations moving through an ordered structure. The first generates disorder, the second "repairs" this, restoring the original structure

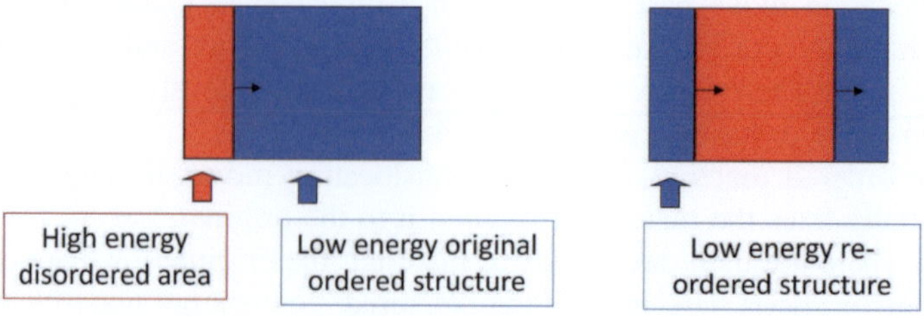

Fig. 4.10 Schematic illustration showing how there is a region of high energy material between two successive dislocations passing through an ordered material

Fig. 4.11 Schematic showing how yield strength behaviour of the constituent phases adds up to the overall behaviour of a nickel based superalloy with the γ/γ' structure. Based on published results [6]

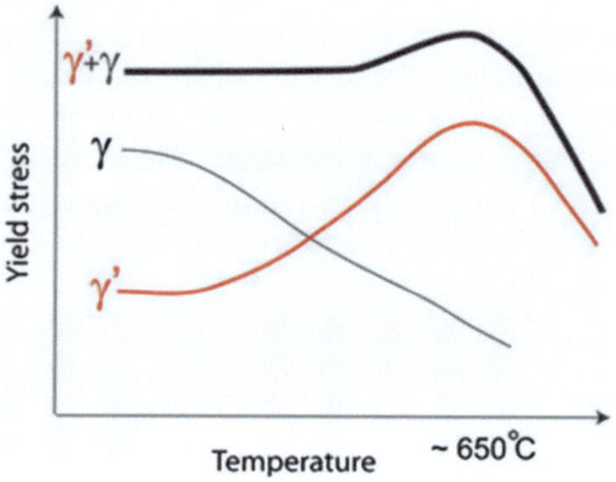

to maintain this separation, and hence the minimum energy. Moving two dislocations at once is more difficult than moving one, this means the ordered gamma prime precipitates have an inherently high yield strength.

The third reason that makes the gamma/gamma prime structure work so well is the yield strength anomaly. This is a feature of the gamma prime precipitate's crystal structure. The yield strength of the gamma, the main nickel matrix, behaves like most metals and decreases with increasing temperature. However, the gamma prime precipitates behave in an anomalous way, their yield strength increases with temperature. This is because higher temperatures thermally activate additional slip systems. At these higher temperatures dislocations can now move in more directions but this means that the dislocations moving in these new directions can now crash into other dislocations. They do this and get stuck, making further dislocation motion more difficult and thereby increasing the yield strength. Together the combination of the gamma prime, where strength increases with temperature, and the gamma, where strength decreases with temperature, gives a gamma/gamma prime structure with a high strength that remains stable to high temperature (Fig. 4.11).

4.3.1.2 Microstructural Development

As well as continuous development and optimisation of alloy composition, microstructural development has been a key factor in the progress of superalloys. This is particularly relevant to turbine blades and has resulted in improved creep resistance.

The first step change was the move to directional solidification. This was achieved by progressively withdrawing the mould from the furnace, ensuring that solidification started from one end of the mould and proceeded along the length of the turbine blade. This results in long grains, parallel to the length of the turbine blade (Fig. 4.12, centre). This solidification pattern removes any transverse grain boundaries. Two of the four mechanisms of creep involve grain boundaries. Removing these grain boundaries immediately improves creep resistance by eliminating the grain boundary diffusion and grain boundary sliding mechanisms of creep.

This microstructural control is pushed further in the generation of single crystal turbine blades. Here all grain boundaries are eliminated, further enhancing creep resistance (Fig. 4.12, right). This is done by adding a grain selector. The process starts with the same directional solidification technique. Once directional solidification is established a grain selector is added in. This is a spiral shaped tube, which allows only one grain to continue to grow. The shape of this is seen in the casting remnant shown in Fig. 4.13, the single crystal turbine blade has been removed from the top of the remnant. Vertical grain boundaries are visible in the bottom part of the remnant, showing that directional solidification was achieved

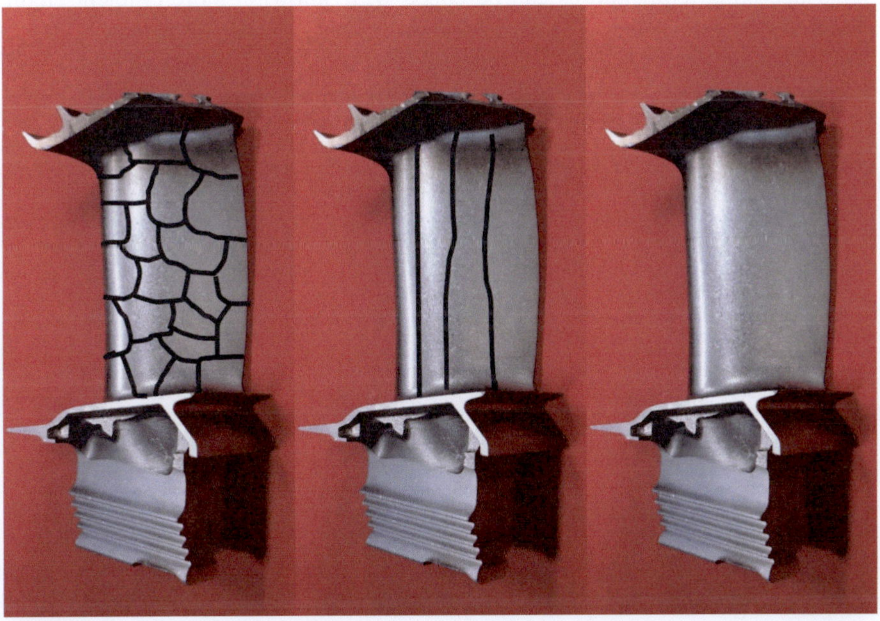

Fig. 4.12 Turbine blade with schematic indication of grain boundaries superposed to illustrate the progression in microstructure from (left) poly crystalline to (centre) direction solidification and (right) single crystal

Fig. 4.13 Casting remnant from pigtail grain selector used to produce single crystal turbine blades. The circular base is approximately 30 mm in diameter. "File:Pigtail from Single Crystal Blade Casting shown with Kennedy Half Dollar for size comparison.jpg" by Lee Langston is licensed under CC BY-SA 4.0

in this part of the material. The resulting improved creep resistance mean that higher operating temperatures can be coped with, allowing more efficient engine operation.

4.4 Test Your Understanding—Questions

Q1 Which of the following is a typical application of a superalloy?
Select one:

a. turbine blade
b. wing skin
c. landing gear

Q2 Which two types of material dominate material usage in engines?
Select one:

a. high strength steel and Ti alloys
b. Ni superalloys and Ti alloys
c. Ti alloys and composites
d. high strength steel and Ni superalloys

Q3 Superalloys have lots of advantages and one key disadvantage, what is that?
Select one:

a. high melting point
b. high density
c. low density

Q4 In the gamma/gamma prime structure what are gamma and gamma prime?
Select one:

a. you can use gamma and gamma prime interchangably as long as you are consistent
b. gamma is the Ni matrix and gamma prime is the alloying elements
c. gamma is the precipitate and gamma prime is the Ni matrix
d. gamma is the Ni matrix and gamma prime is the precipitate

Q5 What feature of the gamma/gamma prime system makes the microstructure of the system stable at high temperatures?
Select one:

a. solid solution strengthening
b. the high density
c. the yield stress anomaly
d. the coherent gamma/gamma prime interfaces

Q6 Which part of the gamma/gamma prime system exhibits the yield stress anomaly?
Select one:

a. gamma prime
b. gamma
c. both gamma and gamma prime
d. the gamma/gamma prime interfaces

Q7 Where do dislocations travel in pairs, and why?
Select one:

a. nowhere, dislocations never travel in pairs
b. in gamma prime as a result of the ordered structure
c. around the edges of gamma prime because of the coherent interfaces
d. in gamma as a result of the ordered structure

4.5 Test Your Understanding—Answers

Q1 Answer a turbine blade

Q2 Answer b Ni superalloys and Ti alloys

Q3 Answer b high density

Q4 Answer d gamma is the Ni matrix and gamma prime is the precipitate

Q5 Answer d the coherent gamma/gamma prime interfaces

Q6 Answer a gamma prime

Q7 Answer b in gamma prime as a result of the ordered structure

Further Reading

Organisations and companies

The Nickel Institute has resources and information on all things nickel, not only superalloys https://nickelinstitute.org/

The International Nickel Study Group also has nickel related resources and links available via its website https://insg.org/

Sandvik's Application Guide to Heat Resistant Super Alloys has a lot of information about using superalloys, including a lot of information relating to machining: https://cdn.sandvik.coromant.cn/files/sitecollectiondocuments/downloads/global/technical%20guides/en-gb/c-2920-034.pdf

https://matmatch.com/learn/material/superalloys overview of superalloys and their applications with relevant links.

Open access articles and research publications

Superalloys for Industry Applications (EBOOK (PDF) ISBN 978–1-83,881–249-2) is an open access peer-reviewed book published in 2018 that has six chapters, each focussing on different aspects of superalloy related research. https://www.intechopen.com/books/5825

Recent Advancements in the Field of Ni-Based Superalloys, Selvaraj, SK; Sundaramali, G, Dev, SJ, Swathish, RS, Karthikeyan, R, Vishaal, KEV, Paramasivam, V Dec 27 2021 Advances in Materials Science and Engineering, Vol 2021 https://doi.org/10.1155/2021/9723450

https://www.theengineer.co.uk/content/in-depth/jewel-in-the-crown-rolls-royce-s-single-crystal-turbine-blade-casting-foundry/ overview of how Rolls-Royce make single crystal superalloy turbine blades, accessed 28/11/23.

References

1. Nickel and the Environment, Policy section of The Nickel Institute website https://nickelinstitute.org/en/policy/nickel-and-the-environment/#:~:text=Nickel%20production%20is%20energy%20intensive,nickel%20during%20production%20by%20far accessed 19/10/23
2. European Commission, Brussels, 3.9.2020, COM(2020) 474 final, Communication from The Commission to The European Parliament, The Council, The European Economic and Social Committee and The Committee of The Regions, Critical Raw Materials Resilience: Charting a Path towards greater Security and Sustainability https://eur-lex.europa.eu/legal-content/EN/TXT/?uri=CELEX:52020DC0474
3. ALLOY 625 - Impressive Past/Significant Presence/Awesome Future, G. D. Smith, D. J. Tillack and S. J. Patel, In Superalloys 718, 623, 706 and Various Derivatives, Ed E.A Loria, TMS (The Minerals, Metals & Materials Society). 2001. https://www.tms.org/Superalloys/10.7449/2001/Superalloys_2001_35_46.pdf
4. https://www.specialmetals.com/documents/technical-bulletins/inconel/inconel-alloy-625.pdf accessed 18/11/23
5. The effect of oxidising thermal exposures on the fatigue properties of a polycrystalline powder metallurgy nickel-based superalloy, D.T.S. Lewis , R.G. Ding , M.T. Whittaker, P.M. Mignanelli, M.C. Hardy, Materials & Design, Volume 189, April 2020, 108529 https://doi.org/10.1016/j.matdes.2020.108529
6. https://www.mdpi.com/metals/metals-06-00037/article_deploy/html/images/metals-06-00037-g002.png accessed 18/11/23

Steel

5

Steel: The bullet points

- Steel is not one single material but a family of many types of steel, with many specific steels of each type
- A little bit of carbon makes a big difference
- Need to know thermomechanical history, not just composition, to know material properties
- Investing some time in learning how to use the iron carbon phase diagram is a route to deeper understanding of steel

5.1 Introduction

Steel is, or rather steels are, the most extensively used metallic engineering material. There are countless applications. Despite its relatively high density it has many applications in transport industries, even aerospace where it is used for engine pylons, landing gear and turbine shafts. Steel dominates the automotive industry. Whilst aluminium alloys and composites are making ground in the rail industry the majority of carriages, rails, wheels and locomotives are steel (Fig. 5.1).

In the marine sector, where the highly corrosive environment cannot be ignored, large marine vessels are steel hulled, with steel superstructures and shipping containers also made from steel (Fig. 5.2). Steel is so widely used because it is strong and stiff and reasonably cheap. It can also be worked and joined relatively easily.

The high density and susceptibility to corrosion are big disadvantages of steel. However, these are outweighed by the advantages, not least the typically low cost (though advanced steels and stainless steels will be higher price), high stiffness and high achievable strengths. High strength steels can have the high specific strengths

K. T. Voisey, *The Engineer's Guide to Materials*,
https://doi.org/10.1007/978-3-031-62937-2_5

Fig. 5.1 Typical usage of steel in rail applications, here steel rails, wheels, bogies and freight carriage bases are seen

Fig. 5.2 Steel shipping containers are widely used and are continuously exposed to weather and the corrosive marine environment

that are so important in transport industries, particularly aerospace, as component mass is inversely proportional to specific strength.

Despite the high density, using steel does not automatically mean heavy components or structures. The high strengths attainable by steel mean that smaller cross-section, and hence lower mass, components can be used. This is particularly true of high strength steels and advanced high strength steels, however the increased strength comes at higher cost.

Fig. 5.3 Backstreet steel workshop in Bangkok

The widespread use of steel means that there is an established manufacturing base. Wherever you are in the world you will find workshops able to work with steel (Fig. 5.3), this is simply not the case for all other materials.

There are many types of steel: mild steel, maraging steels, stainless steels, low carbon steels, TRIP steels, tool steels, interstitial free steels, high strength low alloy steels… With each of these having sub-types, high strength low alloy steels come in a number of varieties including control-rolled, pearlite reduced, microalloyed, acicular ferrite, dual phase… There are various bodies which produce standards for steels, these are generally national institutions each of which has their own system. Comparison tables are widely available to translate between the different steel grading systems [1]. The designator for a specific steel can contain a lot of information, for example BS 7668 S345J0WH. "BS 7668" is the relevant British Standard, "S" indicates that this is a structural steel, "345" is the minimum yield strength in MPa for thicknesses less than 40 mm, "J0" gives information about impact properties, that this is a minimum of 27 Joules at 0 °C and "WH" designates that this is hollow section weather resistant steel.

Despite this, potentially intimidating, list of different steels there are unifying typical characteristics, behaviours, properties and underlying principles. A good starting point is the fact that all steels are based on iron. Iron has a density of 7874 kg m^{-3}, and is the main element present in steel, as a result all steels have densities that are similar to that of iron, typically ranging from 7700 to 7900 kg m^{-3}. Another property that is fairly consistent across all steels is the Young's modulus. The value for iron is 211 GPa and for steels it is typically in the range of 200–211 GPa.

As is seen in other metallic materials, the material property that varies the most is yield strength. This is because yield strength does not depend only on composition, but also on microstructure which itself depends on thermal history and mechanical history.

5.2 Sustainability Considerations Relating to Steel

Steel is highly recyclable and is widely recycled, making it a good fit with the circular economy. In the US any "new" piece of structural steel contains over 90% recycled steel [2]. This extensive use of recycled steel has a big impact on the energy input, and carbon emissions, associated with using steel. With appropriate corrosion protection, steel components can have long lifetimes. There are however huge carbon emissions associated with the steel industry which is responsible for more than 7% of global CO_2 emissions. This is largely due to the massive scale of the steel industry, rather than steel usage itself being in any way inherently unsustainable. Due partly to the simple scale of the industry, there is great scope for decreasing carbon emissions and there is a lot of current activity focussed on using technologies such as green hydrogen to decarbonise steel production. H2 Green Steel's plant in Sweden is one such example, here hydrogen is generated from sustainable electricity and then used in steel production [3]. Other work concentrates on carbon capture and utilisation of the captured carbon dioxide as a raw material for chemical manufacturing and in production of building materials via curing of concrete.

5.3 Applications of Steel

There is an almost endless list of applications of steel. Worldwide, just over a half of all steel, 52%, is used by the construction industry in building and infrastructure applications. Mechanical equipment comes a distant second with 16%, followed by the automotive sector at 12%. Other transport applications make up 5% of steel usage, electrical equipment 3% and domestic appliances 2%. The remaining 10% is attributable to "metal products" [4].

Steel frames are widely used in buildings (Fig. 5.4). Skyscrapers rely on steel frames (Fig. 5.5), the high strength of steel gives it a good load bearing capacity. The greater specific strength of steel compared to concrete means that additional stories in a building will weigh less when built with a steel frame instead of concrete load bearing walls. Another important construction use of steel is in reinforced concrete. Here the tensile strength of the steel reinforcing bars counteracts concrete's poor performance in tension (Fig. 5.6).

Steel is widely used in the automotive sector. Many vehicles have a steel chassis, which exploits the high strength of steel. The good formability and high ductility of steel is made use of in body panels, where the material is rolled down to thin section sheets. Thin sections can raise concerns about corrosion, however

Fig. 5.4 Steel frames are widely used in construction. "Steel construction" by madmack66 is licensed under CC BY 2.0

Fig. 5.5 High rise structures such as the Petronas Towers, Kuala Lumpar, rely on steel framework construction

the widespread use of galvanised steel addresses this and is why manufacturers frequently offer a ten year guarantee against corrosion. The roof, door beams and exhausts are also frequently made from steel. The range of applications within this one sector showcases how adaptable and useful steel is.

Cast iron has been used extensively in automotive engine blocks, due to its good compressive strength, durability and wear resistance. The ease with which it can be cast is also advantageous. It is currently used as the engine block material in the 1 L EcoBoost engine used in the hybrid version of the Ford Puma, though a cast aluminium block is used in the petrol engine version of the same model [5].

Fig. 5.6 Another key use of steel in construction is as reinforcement in reinforced concrete. Here reinforcing bars have been put in place in advance of the pouring of concrete. "Steel reinforcement for an upcoming concrete pour at the Track A approach structure. (CH061A, 11-13-2017)" by MTA C&D—EAST SIDE ACCESS is licensed under CC BY 2.0

The automotive industry exploits high strength steels to decrease total weight. This approach has enabled a 40 kg weight saving to be achieved in the fifth generation Vauxhall Corsa [6]. Hyundai's use of specialised advanced high strength steel in the i30 has resulted in improved rigidity and a 28 kg decrease in weight [7]. The 2019 Ram 1500 truck (Fig. 5.7) has a steel frame that is 98% advanced high strength steel, AHSS, which produced a 45 kg weight saving. AHSS is also used for the truck cab and bed in the same vehicle to enhance overall strength [7]. Volkswagen successfully decreased the CO_2 emissions of its seventh generation Golf by using advanced high strength steel [5]. In 2023 Nissan received an award for their work on increased usage of advanced high strength steel in the Nissan Rogue where improved performance was delivered with minimal mass increase [8]. Toyota and Mercedes-Benz are also increasing their use of AHSS to deliver weight savings and improved structural rigidity [7]. Ford uses ultra high strength boron steel in side impact beams to protect passengers [9], this makes use of the strength and energy absorption characteristics of this steel. The move towards electrification of cars has prioritised the need for weight savings in order to maximise range.

There are also a few aerospace applications of high strength steels. The relatively high density of steel restricts its aerospace applications to where particularly high strength or hardness is required. In applications such as aircraft pylons, engine shafts and landing gear, the advantages of high strength, and consequential decrease in component size, outweigh the disadvantage of high density. Smaller landing gear decrease drag and noise as well as minimising how much payload

Fig. 5.7 The 2019 Ram 1500 has a 98% advanced high strength steel frame which produced a 45 kg weight saving. "2019 Ram 1500 Big Horn Pick-Up" by Crown Star Images is licensed under CC BY 2.0

space is lost to accommodate the retracted landing gear. The 4340 high strength steel is used for the main strut in the Boeing 777 landing gear (Fig. 5.8).

Super Chrome-Molybdenum-Vanadium steel, Super CMV, is used in Rolls-Royce engine shafts. This application exploits the high strength obtainable, 1150 MPa at 450 °C, which is in line with the temperatures experienced by engine shafts [10]. Super CMV also has the creep resistance required to allow it to operate for extended time under load at these temperatures. The usage of high strength steels in this application allows the shaft diameter to be minimised.

Fig. 5.8 Boeing 777 landing gear, main strut is high strength steel. "Boeing 777–200 starboard main gear leg, doors, bogie" by wbaiv is licensed under CC BY-SA 2.0

Fig. 5.9 The high strength steel Aermet 100 is used in the landing gear of this US Navy F/A-18E/F. "The F/A-18 E/F Super Hornet" by Iain A Wanless is licensed under CC BY 2.0. Image shown here has been cropped

This is desirable as it produces a more compact engine. The Trent 1000 engine uses a combination of Super CMV and Aermet 100 for the low pressure turbine shaft [11]. Aermet 100 is an ultra high strength martensitic stainless steel. It was developed to have a reasonable fracture toughness as well as high strength in order to meet the requirements of landing gear applications in the US Navy being used for their F/A-18 E/F fighter jets (Fig. 5.9). It has also been used in landing gear on the F-18, F-22 and F-35 [12]. Slat tracks are another aerospace application of high strength steels. Maraging steel was used for the slat tracks of the Airbus A320 [13].

5.4 How Steel Works

Most people are aware that steel is a mixture of iron and carbon. However, few are aware of just how little carbon is present in steel, and how much of an effect an apparently small amount of carbon can have. Carbon steel is the most widely used type of steel, it has carbon contents of up to 2%. In the world of steel, more than 0.60% carbon in steel is considered a lot, and such steels are regarded as high carbon steel with ultra high carbon steels still only having carbon contents of up to 2%. Cast iron has carbon contents of between 2 and 4%.

The presence of carbon strengthens the material, with steel having a higher strength than pure iron. There are three main ways in which carbon strengthens steel. The first is via solid solution strengthening. This is where the carbon is dissolved in the iron and is present as individual atoms. Carbon is so small that it sits in the gaps, interstices, in the iron crystal structure. However, carbon is actually too big to neatly fit into the interstices. The presence of the carbon distorts the iron lattice, and hence acts as an obstacle to dislocation motion, thereby increasing the yield strength.

The expectation may then be that more carbon means higher strength. This is true, however there is a complication in that the amount of carbon that can be present in solid solution in iron is very limited, at 0.022% carbon. If that limit is exceeded, then something called pearlite will form during cooling. Pearlite is an

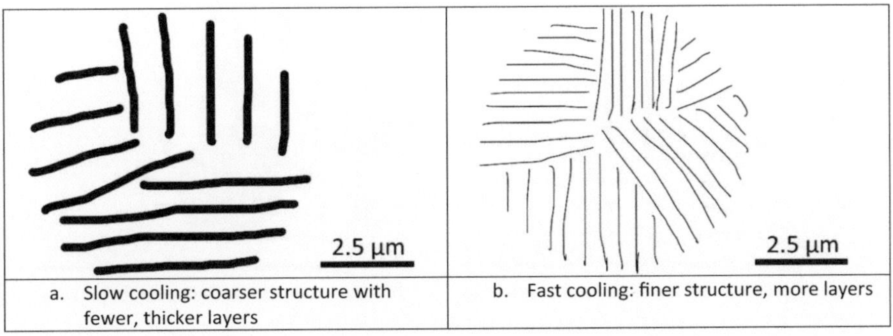

Fig. 5.10 Schematic diagrams of microstructure, circular area is intended to indicate field of view of a microscope. The Impact of cooling rate on the structure of pearlite is seen. Dark stripes are cementite, white material is ferrite. Scale bar is indicative

important constituent of carbon steels. It consists of alternating layers of carbon rich cementite and low carbon iron, ferrite. The structure formed has a distinctive stripey appearance (Fig. 5.10) and pearlite gets its name from the iridescent effect the alternating layers generate on a polished cross-section. Further increasing the carbon content of the steel will increase the proportion of pearlite formed, which will strengthen the steel. This continues up to 0.77% carbon.

Cementite is an iron carbide, Fe_3C, as is typical of carbides, it is a hard, strong but brittle material. The low carbon ferrite is low in strength and ductile. Pearlite behaves as a composite material, with overall properties derived from both constituents. It is stronger than ferrite but tougher, less brittle, than cementite. To form pearlite, the carbon in the steel needs to move. Diffusion is required so that carbon gathers together in the cementite regions, moving out of the ferrite regions. This takes some time to happen. The slower the cooling is the more time there is for diffusion, the carbon can move further and thicker layers form in the pearlite (Fig. 5.10). Rapid cooling generates a finer structure. The interfaces between the layers in pearlite act as obstacles to dislocation motion. The finer structure generated by rapid cooling has more layers and hence more interfaces and a higher yield strength as it contains more obstacles to dislocation motion. This is the same underlying principle that underlies the Hall Petch effect highlighted in the chapter on dislocations (Chap. 12).

Pearlite itself has a fixed composition of 0.77% carbon. The ratio of cementite to iron in pearlite is always the same at 11% cementite and 89% ferrite, approximately 1:8. Depending on the composition of the steel the overall microstructure of the carbon steel with either be pearlite plus additional regions of ferrite or cementite. For carbon contents less than 0.77%, referred to as hypoeutectoid steels, the structure is pearlite plus ferrite (Fig. 5.11a). For carbon contents higher than 0.77%, referred to as hypereutectoid steels, the structure is pearlite plus cementite (Fig. 5.11b). Only at one composition, when the steel composition matches

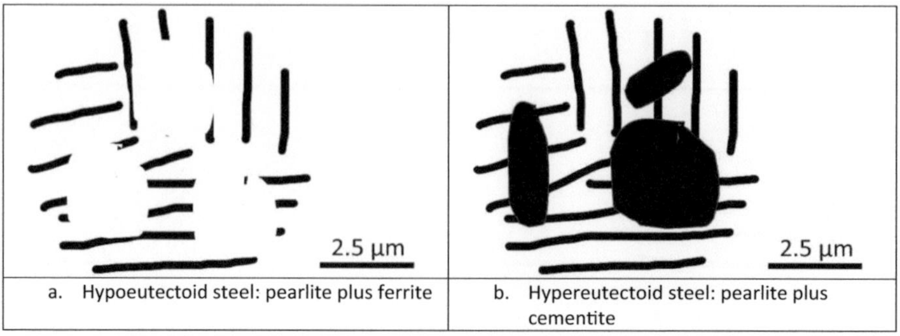

Fig. 5.11 Schematic microstructures of carbon steel showing hypoeutectoid and hypereutectoid steels. Dark regions are cementite, white regions are ferrite. Scale bar is indicative

pearlite's composition of 0.77% carbon, will the steel be 100% pearlite. Increasing the carbon content beyond 0.77% carbon will further strengthen the steel as additional cementite is formed.

The composition of the steel determines how much pearlite can form. The speed of cooling determines the morphology of the pearlite and hence the overall material properties.

The third mechanism by which carbon strengthens steel is by the formation of martensite. This is relevant to steels with at least 0.4% carbon. Martensite is a very hard form of steel. It is generated by very fast cooling which traps carbon in the iron lattice, resulting in extensive lattice distortion. The distorted martensitic structure produced is very effective at obstructing dislocation motion, hence the material has a very high yield strength. Martensite is actually too good at obstructing dislocation motion. This is a problem as the material is too brittle, making it very sensitive to flaws and susceptible to fracture. Martensite often goes through a tempering process in which it is heated to allow some of the carbon to escape from being trapped in the iron. This restores some ductility, making the martensite a useful engineering material.

Martensite formation can be very useful in hardening steels. Flame hardening, induction hardening and carburisation are all processes which deliberately induce martensite. Relevant applications of these processes are given in Chap. 9. The ease with which martensite forms varies with steel composition, higher carbon content makes martensite formation easier, i.e. less severe cooling is required. Alloying elements also affect how easy martensite formation is. Hardenability is the term used to describe how easy martensite formation is. A steel with a high hardenability is easy to harden, i.e. martensite forms easily. Martensite is more challenging to generate in a material with a lower hardenability. Under the same surface cooling conditions martensite will be generated to a greater depth in a higher hardenability steel compared to a steel with a lower hardenability. This is particularly relevant when through thickness hardening is required. Hardenability relates to the depth, not peak value of, hardening.

However, martensite formation can also be undesirable. It can be formed as a side effect of welding. Welding can be regarded as an extreme heat treatment. The near weld material will heat up, and cool down, quickly. This can generate a region of martensite near the weld. Welding can also generate tensile residual stresses in the near weld material. The combination of tensile stresses and brittle martensite leads to cracking problems.

The amount of carbon present in any given steel is fixed during initial production. However, the surface carbon content can be increased by diffusion to enable more effective hardening via martensite production. This is done in the carburisation process, which is explained in Chap. 9.

All materials get more brittle as temperature decreases. In ferritic steels there is a particularly large change in properties which results in significant embrittlement over a relatively small temperature change. This is referred to as the ductile to brittle transition, this is due to the underlying crystal structure but can be influenced by environmental conditions and microstructure. For materials, such as ferritic steels, which have a ductile to brittle transition it is very important to know the temperature at which this occurs and to ensure that this is not within the operating conditions [14].

5.5 Steel and Corrosion

Corrosion is a big concern when working with steel. There are two main approaches used to combatting steel corrosion. The first is adding alloying elements to make the steel corrosion resistant, with stainless steel being an important example of this approach. This can be very effective but does have the big disadvantage of significantly increasing material cost. The other approach is surface engineering where some sort of coating is used to separate the steel from the corrosive environment. This has the advantage of applying expensive, corrosion resistant, material only where it is needed. A relatively cheap underlying steel can be used to provide the required mechanical properties. The issue is that corrosion protection is only provided for as long as the coating remains intact. A brief overview of some approaches to corrosion protection of steel is given below, more information can be found in Chap. 9.

Paint is widely used to protect steel. It works by forming a barrier to water, eliminating contact between the underlying steel and water prevents corrosion. However, any damage to the paint such as scratches that expose the underlying steel will enable corrosion to occur. A more robust approach is to use galvanised steel. Here a zinc layer acts to prevent corrosion of the steel in multiple ways. The zinc coating firstly acts as a simply barrier, again isolating the underlying steel from the corrosive environment. Zinc is a good choice for this as it produces insoluble corrosion products meaning that the zinc itself is inherently corrosion resistant. Galvanised steel is more resistant to breaches than paint. This is due to the cathodic protection effect that arises because zinc has a lower electrode potential than steel. This means that if the zinc is damaged to expose the underlying steel it will be the zinc that corrodes in preference to the steel. In the automotive industry 7.5 μm zinc coatings can give 10–12 years perforation protection. Due to the increased aggressiveness of chloride ion rich marine environments, marine grade galvanised steel has thicker zinc layers.

5.6 Alloying Elements in Steel

Carbon is the most important alloying element in steel but there are numerous other elements that are frequently also added to steel for various reasons. Not everything that is present in a steel is there deliberately, phosphorous and sulphur are often present as undesirable impurities.

A number of alloying elements have more than one role, such as chromium which can be present to enhance corrosion resistance, strengthen via carbide formation or, with vanadium, deoxidises and improve fatigue resistance. Chromium is also an example of where one alloying element works in combination with another in order to improve the alloy. Some of the key alloying elements commonly used in steel are listed below:

Carbide formers, to improve strength: tungsten, chromium
Deoxidises melt: maganese, vanadium (with nickel or chromium).
Enhances corrosion resistance: chromium
Improves strength: molybdenum

Usually, alloying elements in steels are present at lower levels of 0.2–3%, in stainless steels chromium will be present at fairly high levels of at least 11% and typically 16–18%. This ensures that a protective surface layer of chromium oxide, chromia, forms when stainless steel is exposed to the environment. Some more information on stainless steels can be found in Chap. 8.

5.7 High Strength Steels

There is a huge range of strength possible in steels, with some high strength steels achieving ultimate tensile strengths of over 1 GPa (Table 5.1). Various different methods are used to achieve high strengths in steel. Understanding the underlying mechanisms can help to ensure that the material is used appropriately so that the high strength is retained in service.

Firstly, a warning about high strength materials. There is a general trade off between toughness, that is how resistant the material is to crack propagation, and strength. A lot of applications do need high strength materials, but the toughness, or resistance to damage, requirements must not be overlooked. An obvious advantage of high strength materials is that for a fixed load less material is needed to

Table 5.1 Strength and composition of selected high strength steels

	HSLA 100	4340	300 M	Aermet 100	Aermet 310	Super CMV
Yield strength (MPa)	310–620 [15]	470 [16]	1586 [17]	1724 [18]	1900 [19]	1241 [20]
Iron		96	93.4–94.8			Balance
Carbon	0.06	0.37–0.43	0.4–0.46	0.23	0.25	0.23
Chromium	0.45–0.75	0.7–0.9	0.7–0.95	3.0	2.4	3
Manganese	0.75–1.15	0.7	0.65–0.9	0.03		
Molybdenum	0.55–0.65	0.2–0.3	0.3–0.45	1.2	1.4	1.2
Copper	1.45–1.75					
Cobalt				13.4	15.0	13.4
Aluminium				0.004		0–0.8
Nickel	3.35–3.65	1.83	1.65–2	11.1	11.0	11.1
Titanium	0.02			0.013		0–1.3
Silicon	0.4	0.23	1.45–1.8	0.03		
Vanadium	0.03		Min 0.05			
Niobium	0.02–0.06					
Phosphorous	0.02	< 0.035	< 0.035	0.003		
Sulphur	0.006	< 0.04	< 0.04	0.001		

support it. This means that load bearing components made from higher strength materials can be slimmed down compared to lower strength alternatives. This is attractive in terms of reducing mass of components, which is particularly important in the transport applications as it feeds through into fuel cost savings. This makes it worth paying the high cost of high strength steels even in such price sensitive sectors as the automotive industry. However, and this is especially true for steel, it needs to be remembered that thinner section components will have shorter corrosion lifetimes simply because less material needs to be lost before perforation or unacceptable decrease in load carrying capacity occurs.

Despite having low carbon contents of 0.25% or less, high strength low alloy, HSLA, steels have yield strengths greater than 275 MPa. They are many HSLA applications in the automotive industry where they can produce weight savings of 25% compared to carbon steels. Automotive applications of HSLA include wheels, suspension systems, chassis parts, steering parts and door beams. High strength low alloys, HSLA, steels owe their high strengths to a very fine grain size. With the individual crystals, grains, typically 3 μm in size, compared to the hundreds of μm normally seen, there is a huge increase in the number of grain boundaries per unit area. Grain boundaries act as obstacles to dislocation motion. The more grain boundaries there are, the more difficult dislocation motion is and the higher the yield strength is. The impact of this strengthening mechanism can be quantified using the Hall–Petch equation, as detailed in the dislocation chapter (Chap. 12).

The lowest energy state of a crystalline material is when there is a perfect crystal structure. Grain boundaries are a significant disruption to the perfect crystal structure. The high density of grain boundaries that HSLA features therefore represent a higher energy version of the structure and there is thermodynamic driving force to change to a lower energy, large grain size, form of the material. The fine grain size needs to be stabilised using a fine dispersion of precipitates. Niobium, titianium and vanadium carbides and nitrides are used for this. These are sited on the grain boundaries and prevent them from migrating through the material to produce larger grains, thereby allowing the fine grain size to be retained.

4340 is a tempered martensitic steel that is used in aircraft landing gear and also in power transmission systems where it is used in gears and shafts [16]. The main strengthening mechanism is the formation of the martensitic phase. 300 M is a development of 4340. Silicon is added to increase strength and toughness and the addition of vanadium supports grain refinement. Aermet 100 is a secondary hardening martensitic steel. The chromium and molybdenum alloying elements present prevent the steel from losing strength during tempering of the martensite. During tempering of Aermet 100 the martensite decreases in brittleness and strength as usual, but the decrease in strength is countered by the formation of chromium and molybdenum carbides which is triggered by the tempering heat treatment. Aermet 310 is an advance on Aermet 100. Double vacuum melting is used to produce Aermet 310. This advanced manufacturing process comes at a cost but does result in a steel that is free from phosphorous and sulphur impurities, resulting in a higher yield strength than Aermet 100.

It is precipitation strengthening that gives Super CMV the high strength that enables it to be used in gas turbine engine shafts. When compared to AerMet 100, it has superior corrosion resistance, and also has a higher maximum operating temperature due to its better creep resistance [10].

5.8 Cast Iron

For the very high carbon contents encountered in cast irons, 2–4% carbon, graphite can form. The presence of graphite within cast iron makes it self lubricating which can be an advantage. However, the graphite also embrittles the cast steel. This is due to the platelike morphology of the graphite inclusions, in cross-section these look like, and more importantly behave like, small cracks within the material. As a result, cast iron performs well in compression but not in tension. The relatively low melting point of cast iron makes it suitable for casting and it is used for applications such as engine blocks, where the cost, wear resistance, good castability and machinability are advantageous.

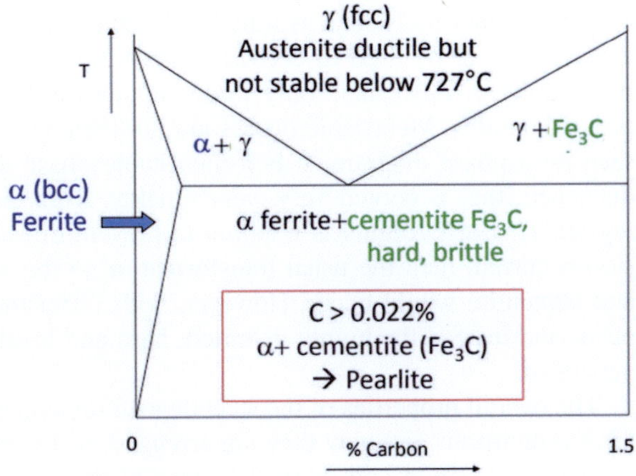

Fig. 5.12 Schematic of the industrially relevant part of the iron carbon phase diagram

5.8.1 The Iron Carbon Phase Diagram

It is worth investing a little time to become familiar with the iron-carbon phase diagram (Fig. 5.12). This contains a lot of information and is an invaluable tool when it comes to understanding the behaviour of steel. Whilst only the iron carbon phase diagram is considered here, this forms a good foundation for understanding the phase diagrams of all steels as these can be regarded as distortions of the main, iron carbon, phase diagram.

The phase diagram has temperature on the y axis, and composition on the x axis. It shows the thermodynamically stable phase, or phases, for any combination of temperature and composition. At least the full phase diagram does. The level of carbon in engineering steels tends to be restricted to less than 1.5%, so it is only the part of the phase diagram that covers the relevant carbon contents of up to about 1.5% which needs to be considered. The phase diagrams of alloy steels can be understood as distorted versions of the iron carbon phase diagram discussed here. This makes understanding the iron carbon phase diagram a useful foundation to understanding the phase diagrams of alloys steels.

In materials, a phase is a physically distinct form of matter. Solid ferrite with solid cementite embedded in it is regarded as a two phase material as it consists of two materials with different compositions and structures, i.e. two physically distinct forms of matter. This also means that there can be solid to solid phase transformations. These solid state transformations feature prominently in the heat treatment of metals and are a key reason why heat treatment can have a significant effect on material properties even when melting does not occur.

Phase diagrams are very useful tools but they do have some important limitations. No information about phase transformation kinetics are included in phase diagrams, other tools such as time temperature transformation diagrams (TTT diagrams) need to be used to determine the time required for any transformation to be completed. Also, it is only thermodynamically stable phases that feature on phase diagrams. Metastable phases are not included. This is why martensite is not seen on a phase diagram. It is a metastable phase that only exists when steel is quenched, that is cooled very, very quickly, from the high temperature austenite region. This fast cooling is required to prevent diffusion from occurring. If diffusion occurred, then the usual transformation to the equilibrium mixture of ferrite and cementite would occur. However, with diffusion being ruled out this cannot occur and instead the highly distorted, hard and brittle martensite phase is instead generated.

The overall properties of the steel depend on which phases are present, their relative proportions and how they are arranged in the material. These in turn depend on the composition and thermal history of the material. For any combination of composition and temperature the phase diagram readily shows if a single phase is present or if the material will be in a two phase region.

Austenite, ferrite, cementite, pearlite and martensite are five key constituents of steels. Understanding each of these materials, their properties and how and why they form gives great insight into the behaviour of a wide range of steels. It allows you to understand how thermal history impacts the properties and behaviour of steel.

	Also known as	Temperature stability	Description	Carbon content	Material properties
Austenite	Gamma	Above 727 °C	Iron carbon solid solution	Up to 2%	Ductile
Ferrite	Alpha	Up to 911 °C	Iron carbon solid solution	Low, up to 0.02%	Less ductile than austenite
Cementite	Iron carbide, Fe_3C	Below 1150 °C	Iron carbide	6.7%	Hard, brittle
Pearlite		Below 727 °C	Alternating layers of cementite and ferrite	0.77%	Intermediate between ferrite and pearlite
Martensite		Requires quenching from above 727 °C		At least 0.4%	Very hard but brittle

Pure iron can exist in two forms: austenite and ferrite. Ferrite is the low temperature form, which is stable up to 911 °C. At higher temperatures the iron will transform into austenite. The difference between these two phases is the crystal structure, and this affects their properties.

The lower temperature form, ferrite, also referred to as alpha iron, α-Fe, has a body centre cubic crystal structure (bcc). The higher temperature form, austenite, also referred to as gamma iron, ϒ-Fe, has a face centred cubic structure (fcc).

This is inherently more ductile than the body centred cubic structure (bcc). Slip systems are ways in which dislocations can move, the fcc structure has more of these than the bcc structure, making it easier for dislocations to move in the fcc structure. A lot of materials are more ductile at higher temperature as the increased thermal activation energy makes dislocation motion easier. In steel the difference is particularly marked due to the step change in behaviour as the material transforms between austenite and ferrite. This is exploited in a lot of manufacturing processes where steel or iron is heated up before forming operations are carried out. At the higher, austenitic, temperatures, the material is much more ductile and hence lower forces are required to shape the material. This means forming can be achieved more easily, with cheaper equipment.

Once carbon starts being added, another important difference between ferrite and austenite becomes apparent: how much carbon can be held in solid solution. For ferrite this is relatively low, a maximum of 0.02% carbon will dissolve in ferrite. For higher levels of carbon, an iron carbide, Fe_3C, cementite will start to form.

Austenite is quite different, the fcc crystal structure is more accommodating. Austenite can have up to 2% carbon in solution before cementite starts to form. That is a big difference, one hundred times as much carbon. Austenite is the high temperature phase, but can be stabilised to lower temperatures in alloy steels, such as austenitic stainless steels including 316, thereby allowing the greater ductility of the phase to be exploited in applications.

Nickel and copper are two alloying elements which act as austenite stabilisers and extend the range of temperature that austenite can exist over. This is useful as it allows the higher ductility of the phase to be exploited in applications. Different alloying elements, such as tungsten and chromium, stabilise the ferrite phase. This is exploited in ferritic stainless steels where the combination of relatively low cost, good ductility and corrosion resistance makes them suitable for applications such as automotive exhausts.

Cementite, like most carbides, is strong but brittle. Pearlite consists of alternating layers of ferrite and cementite. Martensite is a specific form steel. It is generated when high carbon, > 0.4%C, steels are cooled very quickly from above 727 °C.

5.8.2 The Lever Rule and Eutectoid

In addition to knowing which phases are present it is frequently useful to also know their relative proportions. For single phase regions this is straightforward, there is only one phase so 100% of the material is that phase and the phase has the same composition as the overall alloy composition. For two phase regions a simple calculation, using the lever rule, enables the relative proportions of the two phases to be determined.

For a composition C, Fig. 5.13, shows that the material will be within a two phase region at temperatures T1 and T2. The horizontal line drawn at T1 intersects

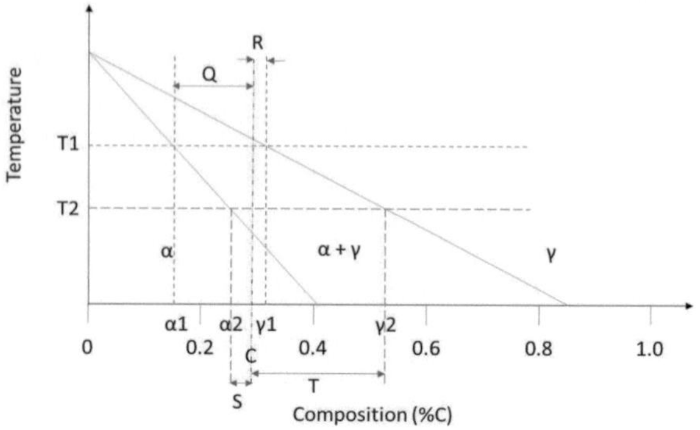

Fig. 5.13 Schematic showing a two phase region of a phase diagram with labels and guidelines for lever rule phase proportion calculations

each edge of the two phase region, the compositions at which these intersections occur are the compositions of the two phases present at this temperature, α1 and γ1.

The lever rule can be used to calculate how much of each phase is present at each temperature. The horizontal line across the two phase region can be split into two segments, Q and R. Q, the segment on the left is between the α composition and the overall composition. R, the segment on the right is between the overall composition and the γ composition. The proportion of γ is determined by Eq. 5.1.

$$\gamma = \frac{Q}{Q + R} \tag{5.1}$$

As there are only two phases present, the proportion of α is simply one minus the proportion of γ. It can be seen that the proportion of the material to the right of the overall composition uses the left hand segment of the horizontal line as the numerator in the fraction, and vice versa. This can be remembered by considering that the majority phase is always the phase which has its composition, at the temperature of interest, closest to the, unchanged, overall composition.

For the lower temperature, T2, it can be seen that when the horizontal line is drawn across the two phase region the intersections have changed. The same two phases are present, but the composition of each phase has changed. The α phase now has a composition of α2 and the γ phase now has a composition of γ2. The proportions of each phase have also changed. It can be seen that at the lower temperature of T2 the, unchanged, overall composition is now closer to the α edge of the two phase region, meaning that this phase will now be the dominant phase. At T2 the lever rule equation for the proportion of γ is as given in Eq. 5.2:

$$\gamma = \frac{S}{S + T} \tag{5.2}$$

It can be seen that as temperature changes within a two phase region both the compositions and proportions of the two phases progressively change. The overall composition however remains unchanged. As a material cools in a two phase region, the compositions of the two phases effectively move down the edges of the two phase region. Their relative proportions constantly change also as must happen to maintain an overall constant material composition. In Fig. 5.13 it can be seen that as cooling continues below T2 the composition of α approaches the overall composition C. At the point when the composition of α reaches C all of the material is in the α phase and cooling will continue within the single phase α region. The point at which the α composition becomes equal to the overall composition is on the boundary between the two phase $\alpha + \gamma$ region and the single phase α region. Carrying out the lever rule here is somewhat pointless but would give the result of the material being 100% α.

Phase diagrams when used with the lever rule can thus be used to determine the evolution of changes to phases within materials as temperatures change. It must however be remembered that phase diagrams represent thermodynamic equilibrium. In reality, thermodynamic equilibrium can take some time to be reached, the thermodynamic equilibrium state is not reached instantaneously. Phase diagrams do not give any information about the length of time required for this to occur. To understand the length of time for any changes to occur, kinetics needs to be taken into account.

The austenite phase can contain, depending on the temperature, up to 2% carbon. For all non-zero carbon contents the material must pass through a two phase region as it cools from the high temperature austenite single phase region. (If the material is pure iron and hence carbon free it will transform directly into ferrite when it cools below 912 °C.)

Compositions with up to 0.77% carbon pass through the austenite and ferrite two phase field, whereas steels with between 0.77 and 2.0% carbon pass through the austenite and cementite (Fe_3C) two phase field. The behaviour of the steel depends on whether the carbon content is higher or lower than 0.77% carbon.

The austenite phase field narrows to reach a point as the temperature decreases and approaches 727 °C (Fig. 5.14). This point corresponds to a composition of 0.77% carbon, meaning that whatever the initial composition, the composition of austenite approaches the same value, 0.77% carbon, as the temperature approaches 727 °C.

0.77% carbon at 727 °C is a special point on the iron carbon phase diagram, this is the eutectoid point. The special thing about a eutectoid point is that here one solid phase transforms into two new, different, solid phases: the 0.77% carbon austenite transforms into ferrite and cementite (Fig. 5.15). The lever rule can be applied to the ferrite and cementite two phase region to show that as the eutectoid point is passed the austenite transforms into a mixture that is 88.8% ferrite and 11.2% cementite, i.e. there is a 8:1 ferrite:cementite ratio.

Steels with carbon contents less than 0.77% carbon are hypoeutectoid steels. Just above 727 °C these consist of 0.77% carbon austenite plus ferrite. Steels with carbon contents above 0.77% carbon are hypereutectoid steels. Just above 727 °C these consist of 0.77% carbon austenite plus cementite. In both cases as the steel cools below 727 °C the 0.77% carbon austenite transforms into a mixture that is 88.8% ferrite and 11.2% cementite, with the additional ferrite or cementite that

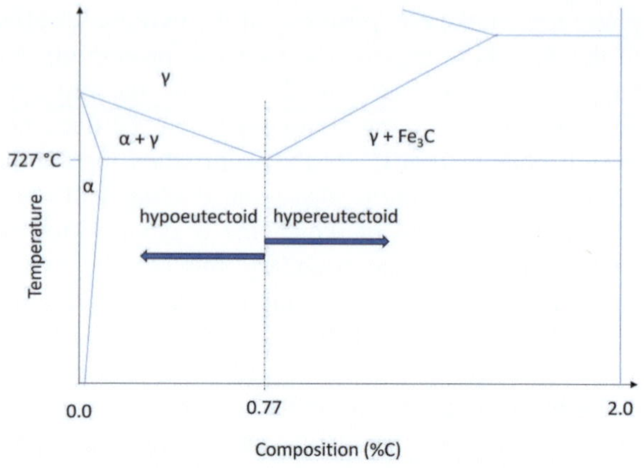

Fig. 5.14 Schematic of the iron carbon phase diagram highlighting hypereutectoid and hypoeutectoid compositions

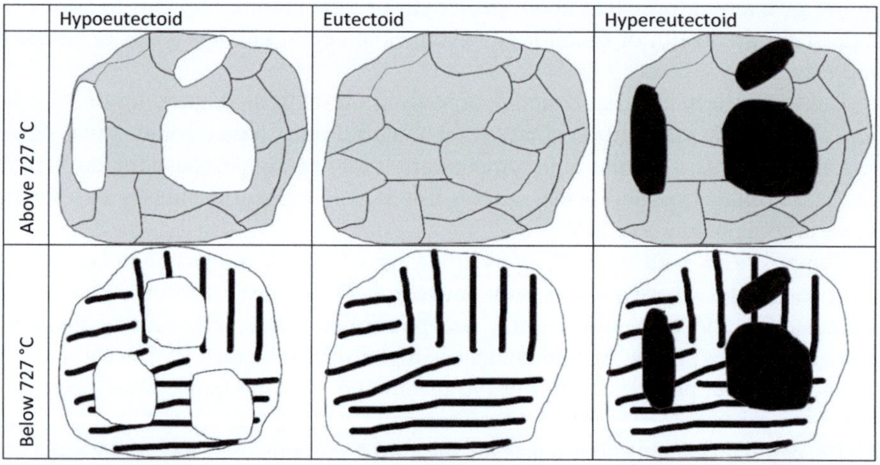

Fig. 5.15 Schematic illustration of hypoeutectoid, eutectoid and hypereutectoid microstructures above and below 727 °C. Grey is used for austenite, γ; white is used for ferrite, α; black is used for cementite, Fe_3C

was present above 727 °C retained. The 8:1 mixture of ferrite and cementite is pearlite which has a distinctive, stripey, microstructure made up from alternating layers of ferrite and cementite. Pearlite can be seen in a wide variety of steels. An overview of this behaviour is given in Fig. 5.15.

Below 727 °C hypoeutectoid steels consist of retained ferrite plus pearlite, whereas hypereutectoid steels consist of retained cementite plus pearlite. Pearlite can be thought of as a composite material, with properties between those of the soft ferrite and strong but brittle cementite. The interfaces between the layers in pearlite act as obstacles to dislocation motion, contributing to the yield strength. For pearlite to form from austenite redistribution of carbon needs to take place.

The carbon that was initially distributed throughout the austenite has to segregate into the carbon rich cementite layers, leaving the ferrite layers almost carbon free. This motion of carbon occurs via diffusion and takes time. The faster the cooling is, the less time there is for diffusion. Fast cooling results in a finer pearlite structure, with thinner layers due to the reduced extent of, and associated decreased distance of, diffusion. This variation in microstructure with cooling rate results in a variation of material properties with cooling rate. The finer microstructure produced by faster cooling results in a higher yield strength due to the larger number of interfaces, obstacles to dislocation motion, per unit area.

The austenite and cementite two phase field looks a little different to the austenite and ferrite two phase field as the right hand side of it, the ferrite side, is a vertical line. The lever rule works in exactly the same way in this two phase field, and the compositions of each of the two phases move along the edges of the phase field as temperature changes. However, the composition of the cementite remains constant as it moves down the vertical right hand edge of the two phase region, reflecting the fixed chemical composition of the iron carbide, cementite Fe_3C.

Using the phase diagram, together with the lever rule, allows the determination of which phases are present, and the composition and proportion of each phase. Doing this for a sequence of different temperatures allows understanding to be gained of the evolution of microstructure, and properties, during cooling.

5.9 Test Your Understanding—Questions

Q1 Give an overview of the use of steel in commercial aeroplanes, you answer should clarify why steel is used despite its relatively high density.

Q2 Give an example of when the formation of martensite in steel is disadvantageous and can lead to problems.

Q3 Sketch and label the microstructure of a HSLA, High Strength Low Alloy steel, and explain how the microstructural features contribute to strengthening.

Q4 A steel contains regions of coarse pearlite. With reference to the iron-carbon phase diagram, state and explain a heat treatment that would convert the coarse pearlite to fine pearlite. Which material properties would be changed?

Q5 What is the name of the commonly seen constituent of steel which has a distinctive stripey microstructure, consisting of alternating layers of ferrite and cementite?

Q6 Compare the properties of ferrite, cementite and pearlite.

Q7 Use the information in Fig. 5.13 to calculate the compositions and proportions of the phases present at T1 and T2 for the 0.3% carbon steel indicated by the line labelled C.

5.10 Test Your Understanding—Answers

Q1 Answer Use of steel not widespread due to high density, restricted to applications requiring high strength—e.g. engine shafts, landing gear, in both these cases high strength steel is used due to its high strength which enables component volume to be minimised.

Q2 Answer There are various possible answer here, the expected answer is during welding where the near weld material can form martensite. This is problematic as welding also results in generation of tensile residual stresses in the same area and having brittle martensite exposed to tensile stresses means that cracking is likely.

Q3 Answer Answer should show a fine grained microstructure, should be labelled as fine grained. The grain boundaries increase strength by acting as obstacles to dislocation motion, (can be quantified via the Hall Petch equation, see Chap. 12). Fine precipitates should also be included and located on the grain boundaries, they contribute to strengthening by stabilising the fine grain structure, without the precipitates the grains would grow, decreasing the amount of obstacles to dislocation motion and hence decreasing yield strength.

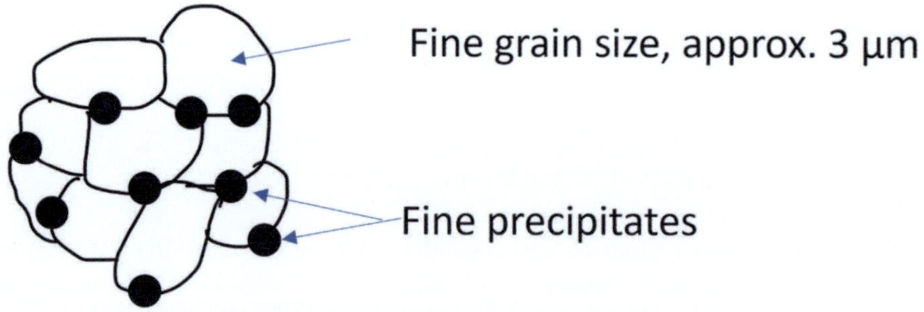

Q4 Answer The steel requires heating into the austenitic region to remove all pearlite, this should then be followed by quick cooling, though not the very very fast cooling referred to as quenching. Quick cooling restricts the time for, and hence distance of, diffusion, resulting in a finer microstructure. This would increase the yield strength and hardness but would also decrease the ductility.

Q5 Answer Pearlite.

Q6 Answer Ferrite is soft and ductile, cementite is hard and brittle.

Q7 Answer The numerical results produced will depend on the accuracy with which numbers are obtained from the figure, there could be some minor variation here that is not a problem as long as the correct method has been used.

T1—2 phases are present α, ferrite, and γ, austenite.

Compositions determined from intercepts with edges of 2 phase region at T1: $\alpha 1 = 0.15$, $\gamma 1 = 0.32$,

C given in question as 0.3, $R = \gamma 1 - C$, $Q = C - \alpha 1$.

Proportion of $\alpha 1 = R/(Q + R) = (0.32–0.3)/(0.32–0.15) = 0.02/0.17 = 12\%$

Proportion of $\gamma 1$, can be calculated using $\gamma 1 = = Q/(Q + R)$, or by noting $\gamma 1 = 1 - \alpha 1 = 88\%$

T2—2 phases are present α, ferrite, and γ, austenite.

Compositions determined from intercepts with edges of 2 phase region at T2: $\alpha 2 = 0.26$, $\gamma 2 = 0.52$,

C given in question as 0.3, $T = \gamma 2 - C$, $S = C - \alpha 2$.

Proportion of $\alpha 2 = T/(S + T) = (0.52–0.3)/(0.52–0.26) = 0.22/0.26 = 85\%$

Proportion of $\gamma 2$, can be calculated using $\gamma 2 = S/(S + T))$, or by noting $\gamma 2 = 1 - \alpha 2 = 15\%$

Further Reading

Organisations and companies

The world steel association has many steel related resources and links on its website: https://worldsteel.org/

The American Iron and Steel Institute has information relating to various aspects of the industry, ranging from information on steel markets to sustainability. https://www.steel.org/

Kayana Engineering's website has various different resources including the detailed blog post "Different Types Of Steel: Benefits, Applications & Fabrication Options", April 1 2022, Graham Dawe, this gives a good overview of different types of steel https://kanyanaengineering.com.au/different-types-of-steel-benefits-applications-fabrication-options/

Tata steel has a lot of information about using steel in different sectors https://www.tatasteeleurope.com/

The British Steel website is another source of information about steel https://britishsteel.co.uk/

There is a lot more to phase diagrams than the brief overview included in this chapter. The tutorials hosted within The University of Cambridge's DoITPoMS freely accessible resource are a good way to start learning more about them. https://www.doitpoms.ac.uk/tlplib/phase-diagrams/index.php

News and open access articles and research

Progress on improving strength-toughness of ultra-high strength martensitic steels for aerospace applications: a review, Jihang Li, Dongping Zhan, Zhouhua Jiang, Huishu Zhang, Yongkun Yang, Yangpeng Zhang, Journal of Materials Research and Technology, Volume 23, March–April 2023, Pages 172–190 https://doi.org/10.1016/j.jmrt.2022.12.177—an open access review.

References

1. https://www.michael-smith-engineers.co.uk/mse/uploads/resources/useful-info/General-Info/MATERIAL-GRADE-COMPARISON-TABLE-for-Web.pdf accessed 11/12/23
2. https://www.aisc.org/why-steel/sustainability/#:~:text=Steel%3A%20The%20Most%20Sustainable%20Choice,become%20steel%20again%20and%20again accessed 28/11/23
3. https://www.h2greensteel.com/who-we-are accessed 28/11/23
4. Global steel usage by sector 2021, Published by Statista Research Department, Mar 7, 2023 https://www.statista.com/statistics/1107721/steel-usage-global-segment/#:~:text=In%202021%2C%20the%20building%20and,percent%20of%20total%20steel%20demand.
5. https://media.ford.com/content/dam/fordmedia/Europe/en/2023/03/PumaST/Ford_Puma-ST_2023_TechSpec_EU.pdf Ford Technical Specifications Ford Puma ST Specifications
6. Vauxhall Reveals All-New Corsa 26 June 2019 https://www.vauxhall.co.uk/experience/vauxhall-news/2019/06/vauxhall-reveals-all-new-corsa.html
7. "Car manufacturers look to steel for stronger, lighter cars" March 2019 | By Desmond Hinton-Beales, World Steel Association https://worldsteel.org/steel-stories/automotive/manufacturers-stronger-lighter-cars-ahss/
8. Nissan Receives Award At 21st Annual Great Designs In Steel Symposium For Innovative Use of Advanced High-Strength Steel May 24, 2023 https://www.steel.org/2023/05/nissan-receives-award-at-21st-annual-great-designs-in-steel-symposium-for-innovative-use-of-advanced-high-strength-steel/
9. https://www.ford.com.ph/engineering/boron-steel/#:~:text=Boron%20steel%20is%20used%20in,on%20their%20size%20and%20usage accessed 28/11/23
10. https://ismswansea.com/materials/steel/ accessed 28/11/23
11. http://www.aengd.org.uk/engd-vacancies/computational-design-and-additive-manufacture-automotive-crash-components-high-performance-sports-cars/corrosion-study-f1e-shaft-steel/ accessed 28/11/23
12. Materials considerations for aerospace applications R.R. Boyer, J.D. Cotton, M. Mohaghegh, and R.E. Schafrik MRS Bulletin • Volume 40 • December 2015 • https://link.springer.com/content/pdf/10.1557/mrs.2015.278.pdf
13. Pfeiffer, H., De Baere, D., Fransens, F., Van der Linden, G., & Wevers, M. (2008). Structural Health Monitoring of Slat Tracks using transient ultrasonic waves. Aircraft Integrated Safety Health Assessment (AISHA), Leuven, Belgium, June 2007. e-Journal of Nondestructive Testing Vol. 13(10). https://www.ndt.net/?id=6967
14. Technical Problem Identification for the Failures of the Liberty Ships, Wei Zhang, *Challenges* 2016, 7(2), 20; https://doi.org/10.3390/challe7020020

15. https://www.metalsusa.com/index.php/hsla100-grade-steel/ accessed 28/11/23
16. https://www.azom.com/article.aspx?ArticleID=6772 accessed 28/11/23
17. https://alloysintl.com/inventory/alloys-steels-supplier/alloy-steel-300m/ accessed 28/11/23
18. https://www.neonickel.com/alloys/stainless-steels/aermet-100/ accessed 28/11/23
19. https://www.matweb.com/search/datasheet.aspx?matguid=0c4383b4e824496eb3721b81646 54fc7&ckck=1 accessed 28/11/23
20. Hyde TR. *Development of a representative specimen for fretting fatigue of spline joint couplings.* PhD thesis: University of Nottingham, 2002.

6 Polymers and Polymer Composites

The bullet points-polymers

- Key properties: low density, high formability
- Knowing if a particular polymer is a thermoset or thermoplastic gives insight into properties and which manufacturing processes are suitable
- Relatively low typical maximum operating temperatures of approx. 100 °C (though specific polymers have higher temperature capacities, such as PEEK which can operate at up to 260 °C)
- Properties depend on processing, not just chemistry
- Properties are highly temperature dependent

The bullet points-polymer composites

- Properties can be tailored by type, amount and architecture of reinforcement, for fibre reinforced composites this can lead to highly directional properties
- High temperature operation limited by polymer matrix
- Challenges related to recycling
- Material is generally made at the same time as the component

6.1 Introduction

Polymers and polymer composites share the underlying properties of low density and fairly low maximum operating temperatures. Due to polymers forming the matrix for polymer composites there is extensive overlap in relevant topics

K. T. Voisey, *The Engineer's Guide to Materials*,
https://doi.org/10.1007/978-3-031-62937-2_6

between these two groups of materials. However, there are significant differences in properties, cost and usage. Non-reinforced polymers are routinely used for mass produced products that do not require high strength and are so cheap that they are frequently regarded as disposable. Fibre reinforced polymer composites in contrast have good specific properties. The market is dominated by glass fibre reinforced composites, these can be used for load bearing structures that are continuously exposed to the elements, such as glass fibre boats. Carbon fibre reinforced polymers are more costly high end, superior, engineering materials which as well as being used in many demanding applications are also made use of as a premium, status symbol, material to increase the consumer appeal of various products.

An important engineering application of polymers is as the matrix in carbon fibre reinforced polymer, CFRP, composites. The high specific stiffness and strength of fibre reinforced polymer, FRP, composites are their headline advantageous properties, the associated reduction in weight and improved fuel efficiency make them attractive to all transport industries. Other advantages include the ability to tailor directional properties and the generally good corrosion resistance, though there can be issues with solvents, including water. The extensive use of FRP in transport applications, ranging from aircraft wings to F1 components, where fluctuating stresses are frequently encountered, means that fatigue is a concern. This is also true of wind turbine blades, another big market for FRP. FRP are fatigue resistant but there is limited data on their behaviour and fatigue related damage mechanisms in FRP can be more complex and difficult to detect than in metallic materials. This is compounded by complications arising from variability in properties which can be caused by non-uniform fibre placement due to issues in resin infiltration leading to resin rich and resin poor areas.

The polymer matrix limits the maximum operating temperature, this is typically about 100 °C for the standard CFRP epoxy matrix but can be pushed higher, as high as 350 °C by use of specialised resins and curing agents [1]. This is still significantly lower than the thermal capability of the carbon fibres themselves, which can usually survive up to 600 °C without problems. The intimate mixing of different materials which is a fundamental feature of FRPs makes them inherently difficult to recycle due to the challenges of separating the materials. The relatively high cost of CFRP also needs to be considered. A lot of this is simply due to the fact that the manufacturing base is still being established. Using CFRP can result in life cycle cost savings. This is achieved not only in decreasing fuel costs but also via part integration which decreases total manufacturing costs by removing several assembly steps.

Polymer composites have vastly superior properties compared to non-reinforced polymers, however there are also numerous applications of non-reinforced polymers. These do, generally, have fairly low strengths and stiffnesses, particularly compared to metals, but their low density and ease of forming lead to many applications. There are many uses of polymers as clips, fasteners, hinges connectors etc., exploiting the ease of moulding. Their low electrical conductivity and, for some, the ability to be transparent, also lead to additional applications. There is extensive use in vehicle interiors, lighting and window applications as well as electrical wiring insulation.

A warning about terminology relating to composites. There are various, very different, forms of composite material. These include ceramic matrix composites, metal matrix composites and fibre reinforced composites which itself can be subdivided into carbon, glass and aramid fibre composites. However, within each field the word composite can be used to describe the material. In any conversation or reading about composites it is definitely worth double checking exactly which type of composite is being considered. In addition to this, CFRP is sometimes simply referred to as carbon or as black metal.

6.2 Sustainability Considerations Relating to Polymers and Polymer Composites

There are a number of significant sustainability concerns relating to polymers and polymer composites. The very low cost of plastic has led to countless applications of single use, disposable plastic, particularly in packaging. When combined with the longevity of the material this leads to serious, well documented, environmental issues due to nondegradable polymer waste. Various initiatives exist to address this, with polymer recycling now widely available and a growing number of countries and companies committing to decrease single use plastics and increase utilisation of recycled plastic. Evian has declared that it is on a mission to make their plastic bottles from 100% recycled PET by 2025 [2]. Unilever have set 2025 as the target for their goals of using 25% recycled plastic in their packaging and ensuring that all their plastic packaging is either reusable, recyclable or compostable [3]. Thermoplastics, such as PET, will melt when heated allowing them to be moulded into new shapes. Thermosets however will not melt but will char or thermally degrade meaning that recycling of them is restricted to applications such as using shredded thermoset waste as a filler.

The most widely used polymer matrix in CFRP is thermoset epoxy. There is active research in expanding the use of thermoplastic matrices. The intimate mixture of materials that is an inherent feature of polymer composites complicates their recycling. The growing environmental pressure for industry to operate sustainably makes the recycling challenges an issue. In the automotive industry legislation relating to material reuse and recycling feeds into material selection decisions. The end of life vehicle, ELV, requirements for the UK, and Europe, require 85%, by weight, recycling [4]. There has been research on stripping the matrix from the fibres and reusing the fibres, albeit in less demanding applications as the recycled fibres are much shorter and less robust than virgin fibres. Whilst recycling of polymer composites is more complicated than recycling of polymers, the higher value of polymer composites changes the challenge. Polymer composite components are not single use, disposable, items.

Macroscopic pieces of plastic form one part of the waste problem, and microplastics another. The term "microplastics" applies to any plastic fragment

5 mm or less in size. The impact of microplastics is not yet fully understood. What is known is that they can be found almost everywhere, from the top of Everest to Antarctica [5]. Their small scale means that they can accumulate in seafood and then enter the food chain where their impact is the subject of active research. Another issue is that polymers are generally derived from petrochemicals [6], making them inherently unsustainable. This is not to do with carbon emissions as the carbon is locked within the polymer rather than being released into the atmosphere. The concern is that the fossil fuel resource is being depleted far more rapidly than it can be replaced. Alternative precursors are available, various plant based polymers are now commercially available. These include PLA, polylactic acid and CPLA the higher strength crystallised polylactic acid form which uses lactic acid as a precursor. Such plant based polymers are also designed to be biodegradable. Bio-based polyethylene, polypropylene and PET are also available, however cost challenges need to be addressed to enable large scale commercial application [7].

Despite the recycling issues, which are significant, the very good specific properties and associated low mass of polymer composite components mean that they can positively contribute to overall sustainability by decreasing fuel burn compared to other materials.

6.3 Applications of Polymers and Polymer Composites

6.3.1 Polymers

There are plenty of applications for non-reinforced polymers, though due to their less impressive mechanical properties these tend not to be structural applications. Polymers' typical characteristics of low density, low cost and good formability make them attractive for many applications. Consider the passenger cabin of any commercial aircraft (Fig. 6.1). Pretty much everything you can see, the interior panels, carpeting, seat cushions and window liners will be made from polymers. There will also be many unseen components such as brackets, hinges, connectors and the insulation on electrical wiring. There are similar applications in the automotive, rail and marine industries. More widely there are many packaging applications, not least the extensive use of the thermoplastic PET in drinks bottles.

Non-reinforced polymers can be transparent, which leads to lighting and window applications. Aircraft windows (Fig. 6.2) are made from multiple layers of polymer. Stretched acrylic is the key material used in many different aircraft, including the Boeing 787 Dreamliner. This is a more highly impact resistant version of acrylic. Multiple panes are used: the outer pane carries the load, the inner pane is present as a back up in case of any failure of the outer pane. There is also an additional, passenger facing, pane known as the scratch pane which is present to protect the inner pane from damage.

Polycarbonate and PMMA, polymethyl methacrylate, are both widely used in automotive headlights and interior lighting applications. The high optical transparency, low mass and low cost are key to these applications. The same materials

Fig. 6.1 Interior of a Japan Airways Boeing 777, image courtesy of Helen Ramsay

Fig. 6.2 Fuselage window in an Airbus A320, image courtesy of Charlotte Voisey

Fig. 6.3 A "hot air balloons at sunset" by erink_photography is licensed under CC BY 2.0. To view a copy of this license, visit https://creativecommons.org/licenses/by/2.0/?ref=openverse

are also used in automotive windows, though these are restricted to small side windows rather than the windscreen due to the higher strengths needed for windscreens.

Drawn polymers open up a different range of applications. Here the long chain molecules drawn out into an extended, straightened, configuration to allow the strength of the carbon–carbon bonds that form the backbone of the molecules to be exploited. Drawn polymers can be used to form very high strength fibres which can then be woven to generate high strength fabrics. Hot air balloons and airships use such materials for their envelopes (Figs. 6.3 and 6.4). These are made of multiple layers, typically using high strength fibres such as Zylon® or Vectran™ for the load bearing layers with films of polyurethane or Mylar® then providing the gas barrier [8, 9].

6.3.2 Polymer Composites

The high specific strength and stiffness of CFRP give it many applications across the transport industries where the resultant lower mass components and structures reduce fuel consumption. This is particularly important in aerospace, where the usage has significantly increased over recent years. The Boeing 777, which first flew in 1994, had only 12 weight % whereas the more recent Boeing 787 Dreamliner, in commercial service since 2011, has 50 weight % of polymer composites.

Fig. 6.4 "Hybrid Air Vehicles Airlander 10 'G-PHRG'" by HawkeyeUK—Lest We Forget is licensed under CC BY-SA 2.0. To view a copy of this license, visit https://creativecommons.org/licenses/by-sa/2.0/?ref=openverse

CFRP usage in the Boeing 787 Dreamliner includes significant large load bearing structural components such as the carbon laminate fuselage, wing and tail surfaces. The same trend is seen in Airbus's aircraft. The Airbus 380, which first flew in 2005 had approximately 25 weight % of polymer composites. The Airbus 350 (Fig. 6.5), in service since 2015, has 53% [10]. The 32 m long wings of an Airbus A350 are some of the largest CFRP structures manufactured [11]. CFRP is also used for the A350's centre wing box, fuselage skin panels, tail cone and keel beam.

There are numerous automotive applications, particularly in the high end and motorsport vehicles part of the sector: the whole McLaren range uses CFRP chassis/monocoques (Fig. 6.6); Aston Martin's Valkyrie Amr Pro uses CFRP for the body and some suspension components; the Koenigsegg Agera RS uses both CFRP and Kevlar® in the bodywork; the Zenvo TSR-S uses CFRP alongside other composite materials in the bodywork; Caterham use a CFRP dashboard (Fig. 6.7). In each case the priority has been to minimise vehicle mass to maximise performance. The high specific strength and stiffness of CFRP is key to making this possible. The Audi R8 Spyder convertible top compartment lid is CFRP as are the sideblades on the GT version.

There is extensive use of CFRP in Formula 1 with 85%, the vast majority, of the structure being CFRP [12]. Inclusion of CFRP in front, rear and side crash

Fig. 6.5 An Airbus A350, more than 50 weight % polymer composite, including CFRP wings and key structural components. Image courtesy of Charlotte Voisey

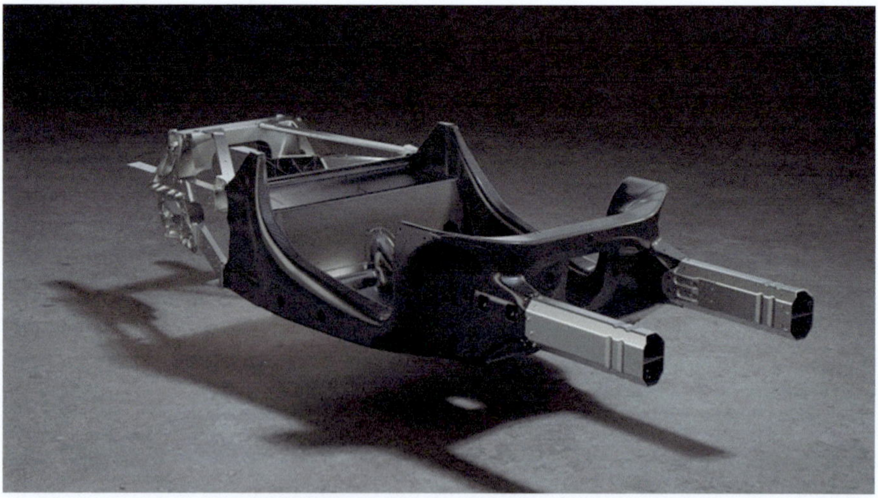

Fig. 6.6 CFRP chassis for the McLaren hybrid Artura, crash cans and rear frame have been added. Image courtesy of McLaren Automotive Ltd

Fig. 6.7 Interior of a Caterham showing CFRP usage in the dashboard

structures exploits the high energy absorbing capacity that comes from the fibre pullout mechanism.

The electric car sector has particular interest in CFRP. Maximising travel distance between charging is a key priority, the lower mass structures made possible by the good specific properties of CFRP help in increasing vehicle range. BMW is one company that have invested significantly in this area [13]. BMW used CFRP extensively in the i3 and i8 models and are now incorporating it in the multi-material design approach being used in the BMW iX. This has a CFRP interior frame, CFRP pillars and roof frame combined with an aluminium alloy floor assembly and steel exterior side frames. This multi-material design delivers increased body stiffness and weight saving [13].

The lack of through thickness electrical conductivity means that CFRP has advantages for defence applications regarding stealth as it decreases the radar signature. Approximately 70% of the surface area of the Eurofighter is CFRP.

Despite the low maximum operating temperature, CFRP is used in aerospace engines. The low maximum operating temperature restricts CFRP to the front, cold, end of the engine. CFRP is used in some fan blades, the large blades right at the front of the engine. The use of CFRP in this application is growing. The driver here is the weight saving, and consequent fuel cost saving, that can be achieved compared to Ti alloy fan blades. GE, General Electric, have been using CFRP blades since 1995. Rolls-Royce's new CTi fanblades flew for the first time in 2014 and are set to be used in their UltraFan demonstrator engine. Usage of fibre reinforced polymer composite in both the fan blades and fan case will deliver

Fig. 6.8 GE90 engine with CFRP fan blades, image courtesy of Antony Smith

a 700 kg weight saving on a two engine aircraft [14]. The "Ti" in CTi refers to the Ti alloy strip added to the leading edge of the fanblade, this exploits the increased ductility of the Ti alloy to endow the blade with impact resistance where it is needed. GE use a similar combination of Ti and CFRP for their GE90 fan blades. An array of these is shown in Fig. 6.8, the metallic Ti leading edge stands out clearly from the black CFRP blade body. A single GE90 fanblade is mounted and on display in the design collection at New York's Museum of Modern Art. As well as celebrating this specific application the display showcases the complex geometry that can be achieved in CFRP components. Some of this 3D curvature is visible in the lower half of Fig. 6.8.

Glass fibre reinforced polymer, GFRP, composites are significantly cheaper, but have significantly lower specific stiffnesses, than the more expensive CFRP. In aerospace GFRP is used for radomes, which house communication systems and need to be transparent to electromagnetic waves. Within the fuselage, interior panels and fittings may be made from glass fibre reinforced polymers, preferred to metal components due to the weight saving and having superior properties when compared to non-reinforced polymers.

Rail applications include seat shells and carriage doors, which deliver some weight reduction compared to metallic alternatives [15]. Fibre reinforced foamed urethane has been used for railway sleepers for about 50 years despite issues with cost [15]. Various potential structural applications for FRP in the rail industry have also been identified, such as bogies, however despite a lot of research and design of prototypes this application has not yet been commercially implemented due to cost and technical challenges [15]. GFRPs have many marine applications. The non-metallic nature of the material, coupled with its good specific properties and ability to withstand extended contact with water made GFRP the material of choice for minesweepers such as HMS Wilton (Fig. 6.9) [16]. This ship was

Fig. 6.9 HMS Wilton, the first GFRP minesweeper. "HMS Wilton—geograph.org.uk—1289223" by William is licensed under CC BY-SA 2.0

launched in 1972 and is now repurposed as the clubhouse of the Essex Yacht Club, Leigh-on-Sea, Essex. GFRP is used extensively in small ships and boats, with the vast majority of hulls for ships less than 20 m being GFP [16]. Hovercrafts and catamarans also frequently use GFRP. There are many other marine applications such as topside structures, hatch covers and radomes as well as powerboats [16].

There are similar automotive applications for glass fibre composite where it is used in non-structural applications such as bonnets, door panels, sunshades and bumpers. Such applications often use a form of fibre composite called SMC, this is sheet moulding compound which is a short fibre composite which is moulded to shape via application of heat and pressure. The structural parts of the Airlander (Fig. 6.5) uses CFRP and GFRP [17], again due to their good specific properties.

Wind turbine blades (Fig. 6.10) are frequently made from GFRP [18], with CFRP also being used. This application exploits the high specific stiffness and good corrosion resistance of these materials. The high specific properties of GFRP and CFRP enable the construction of large blades, these are typically 35–45 m in length, no other material comes close to being suitable.

GFRP is also widely used in applications such as bath tubs, tanks and roofing. The corrosion resistance, high strength and high specific strengths are key properties here. GFRP can withstand continues exposure to weather. This durability, combined with the ease with which GFRP is moulded is exploited in many customised exterior applications such as kiosks, architectural features and statues

Fig. 6.10 GFRP and CFRP are the only materials with high enough specific properties to be viable options for the formation of wind turbine rotors, large components that are continuously exposed to the weather. "A few wind turbines" by vaxomatic is licensed under CC BY 2.0

(Fig. 6.11) where the low mass resulting from the high specific strength is also an advantage.

6.3.2.1 How Non-reinforced Polymers Work

Polymers famously consist of long chain molecules, with each link of the chain being repeating unit referred to as a monomer. The number of monomers in a polymer molecule can be huge, over ten thousand is not unusual. The molecular length does have some impact on properties, with there being some degradation of properties as average molecular length gets shorter. This is due to the extra space in the structure associated with the ends of the molecules, and shorter molecules have more ends per unit volume. The chemistry of that repeating unit has a massive impact on the properties of the overall polymer, but that is fixed during initial molecule formation and is generally unchanged during material lifetime. Chemistry is not the only other factor. The material properties also depend on the shape adopted by the molecule, and this can be determined by how the material is processed. This means that chemically identical polymers can behave differently.

The most extreme molecular configurations that a polymer molecule can adopt are (1) Amorphous and (2) Drawn (Fig. 6.12).

In the amorphous configuration (Fig. 6.12a) the molecule is, a random, spaghetti-like tangle. This happens when there are only weak bonds between molecules, and between different segments of the same molecule. Polymer

Fig. 6.11 The Headington Shark, Oxford, UK, officially known as "Untitled 1986". This example of the use of GFRP in a customised application has been in place for almost 40 years, with the GFRP itself not requiring any repairs despite constant exposure to the weather. GFRP was selected for use due to it being easy to sculpt, its good weathering resistance, strength, low density and low cost. More information on this specific application can be found at www.headingtonshark.com. Image courtesy of Julian Bond

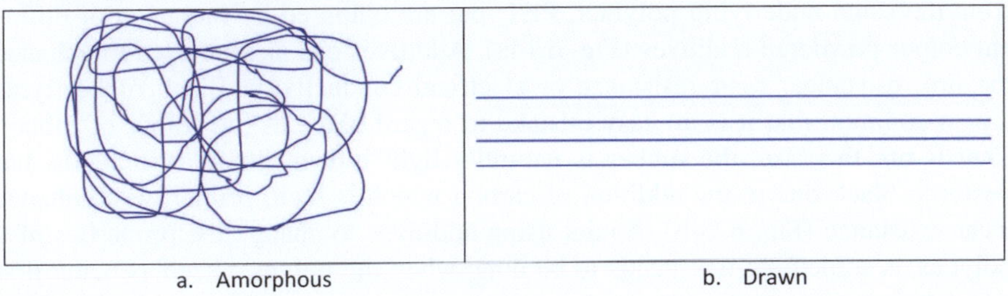

Fig. 6.12 Schematic diagram showing **a** amorphous and **b** drawn polymer molecular configurations

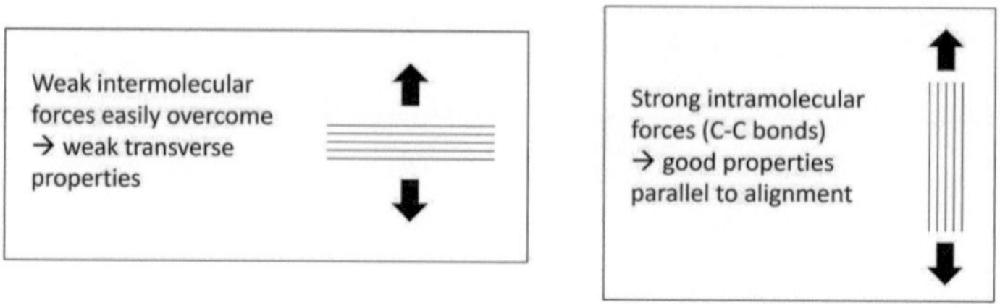

Fig. 6.13 Schematic diagrams illustrating how weak intermolecular and strong intramolecular forces result in directional properties of drawn polymers

molecules are so long that they can cross over themselves and interact with another part of themselves as if it was a different molecule. This means that there is no preferential orientation, hence a random configuration with no alignment results. Such polymers have isotropic properties.

The other extreme is a drawn molecule (Fig. 6.12b). This occurs when the material has been drawn out and the molecules end up straightened out and aligned in one direction. This is analogous to a fibre composite, the drawn polymer can have impressive strength and stiffness parallel to the length of the molecules, with significantly poorer properties in the transverse direction. This is due to the different bonding in different directions (Fig. 6.13). Any force applied parallel to the molecule length will be pulling against the strong carbon–carbon bonds that form the backbone of the molecule. In contrast, transverse stresses will be acting against the, much weaker, Van der Waals bonds between the molecules.

The same polymer can exist in either amorphous or drawn form depending on how it has been processed. Hence the properties also depend on processing. The form developed during processing may be changed by future use of the material. The impact of such changes needs to be understood to ensure appropriate use of the material.

Changes to the overall polymer properties can be achieved by the use of additives. The different colour lids for plastic soft drinks bottles are largely made from the same underlying polymer, PET, but are coloured by inclusion of different colour powdered additives (Fig. 6.14a). Additives can also be used to enhance the fire resistance, wear resistance or electrical conductivity. Black rubber tyres are so common that it is an easy mistake to regard black as the colour of rubber. That is not the case, the rubber is naturally light brown, like elastic bands, but becomes black due to the addition of carbon which is there primarily to enhance wear resistance (Fig. 6.14b). Again, using additives to change the properties of a polymer is something that needs to be done when the polymer itself is being first generated.

The difference in properties that is achieved via changing the configuration of the molecules between random and drawn is not irreversible. Thermodynamics drives polymers to take up the random, spaghetti shape wherever possible, i.e. when there is enough thermal activation energy. As illustrated in Fig. 6.15, this can result in distortion. These disposable coffee cups are manufactured from a

a b

Fig. 6.14 Examples of the use of additives in polymers **a** to change colour in bottle tops, **b** to improve wear resistance in rubber tyres

disc of polystyrene. This is heated and pulled out to form the cup shape. The heat allows the molecules to slide past each other under an applied force. As the molecules move they become extended, straightening out in the direction of flow parallel with the sides of the cup. This configuration gets frozen in as the cups cool rapidly after being formed. However, subsequent heating can provide the thermal activation energy needed to remobilise the molecules, allowing them to move and change their shape back toward the favoured spaghetti-like shape. As seen in the example of the coffee cups, directional alignment will result in directional shrinkage, producing overall distortion of the shape. Minor distortion of a disposable coffee cup is not a large problem, however directional shrinkage of a lid or closure could have greater impact.

Any polymer manufacturing process in which molecules flow can result in alignment and subsequent heat induced distortion. These include injection moulding and blow moulding. Distortion issues can be minimsed if the polymer cools slowly in the mould, however this decreases production rates.

Thermosets/Thermoplastics

Thermosets and thermoplastics are the two main types of polymer. Knowing which category any particular polymer falls into gives insight into its properties, which manufacturing processes are available, recyclability and behaviour on exposure to high temperature.

Thermoplastics are characterised by weak bonds between molecules. If they are heated these weak bonds are easily overcome and the molecules will slide past each other, allowing the overall shape to be changed, hence the "plastic" part of the name. In contrast, thermosets have strong permanent chemical bonds between molecules. Thermosets do not melt, but instead char or thermally degrade when heated excessively. This means that, once the permanent bonds are in place, the shape of a thermoset cannot be changed which gives particular challenges in recycling. However, the stronger bonds of thermosets mean that they tend to be

Fig. 6.15 Two polystyrene disposable coffee cups. The cup on the left is unused, the polymer molecules will have become aligned as schematically indicated during the manufacturing process. The cup on the right was submerged in boiling water for an extended period of time allowing thermal activation energy for the molecules to move and start to revert to the favoured spaghetti like configuration, resulting in distortion

stronger, stiffer and harder, but more brittle, than thermoplastics. For thermosets the formation of the strong permanent chemical bonds occurs during curing. This is the same curing process that CFRP undergoes, the epoxy matrix being a thermoset.

6.3.2.2 How Carbon Fibre Polymer Composites Work

The Boeing 787 fuselage is a huge CFRP composite component. Like the vast majority of CFRP this makes use of epoxy as the matrix. Epoxy requires curing, this is the process where permanent chemical bonds are formed between epoxy molecules, transforming the tacky, flexible material into a rigid matrix which holds the reinforcing fibres in place. Curing requires thermal activation. This is achieved via autoclaves which are essentially pressurised ovens used to heat pre-preg composite parts to induce curing. (Pre-preg is a form of polymer composite where sheets of fibres have been pre-impregnated with the polymer matrix. This is a relatively easy to handle form of the material which ensures good matrix infiltration of the fibres and good fibre alignment). Large parts mean large autoclaves. Vought Aircraft Industries, who manufacture the Boeing 787's rear fuselage have an autoclave which is over 23 m long and 9 m in diameter.

The overall properties of a composite depend both on how much of the reinforcing material is present and how that is distributed within the matrix. The ideal composite would have reinforcement exactly, and only, where needed, meaning very efficient usage of material. However, in practice, the challenges of reinforcement placement would be overly complex and complicated.

All polymer composites combine a reinforcement, which has desirable properties such as high strength and/or high stiffness. This is combined with a low density polymer matrix, the material in which the reinforcement is held, to generate a composite material which combines the advantages of both constituent materials. The

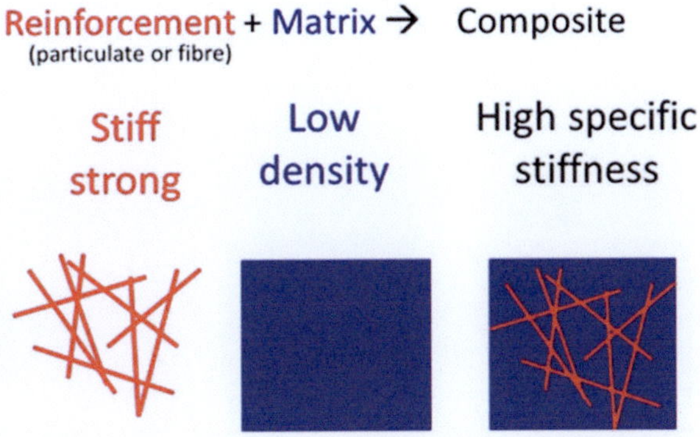

Fig. 6.16 Combination of reinforcement with a matrix material to form a composite

matrix holds the reinforcement in place and, via the matrix/reinforcement interface, transmits applied stresses to the reinforcement (Fig. 6.16).

Reinforcement distribution and the interface between the reinforcement and the matrix are key when working with composite materials. The expected composite properties will only be delivered if the reinforcement is located, and dispersed, as intended. In most cases this means uniform distribution of reinforcement. The advantageous properties of the reinforcement will only be successfully exploited if there is effective load transfer from the matrix to the reinforcing material, and this depends on the quality of the reinforcement-matrix interface.

For the Boeing 787 fuselage, fibre distribution is controlled by using pre-preg material. This consists of sheets, or strips, of fibres pre-impregnated with resin. Prior to curing the pre-preg material is flexible and can be wrapped around the mould to achieve the desired material thickness and fibre alignment. This is all done at an impressively large scale for this application. The use of pre-preg means that the reinforcement is already appropriately distributed in the polymer prior to curing. Pre-preg is a very convenient way of controlling fibre distribution.

Another approach to fibre placement is using dry fibre fabric. The fabric uses tows, which can consist of thousands of individual carbon fibres, as yarn which is then woven in fabric sheets. The sheets are flexible and can be cut and draped over complex moulds. A lot of knowledge on different weave patterns has been directly transferred from the textile industry. The polymer then infiltrates the fibres in a process referred to as Resin Transfer Moulding (RTM). This has been used for applications such as the A320 spoiler assembly, CFRP spoilers are used throughout the A320 family of aircraft, including the A321 (Fig. 6.17) [19]. The low density and high specific stiffness are relevant to ensure effective control of the aerodynamic surfaces without adding significant mass to the aircraft.

This control of fibre alignment allows optimal use of the reinforcing material. Maximum strength and stiffness is achieved when all fibres are aligned and stresses are applied parallel to the fibre alignment. However, transverse properties are significantly lower. If strength in more than one direction is required this can be achieved by having different fibre orientations in different layers of the composite.

It may be thought that uniform, isotropic, properties may be regarded as desirable however a greater fibre fraction, which would be more expensive, would be

Fig. 6.17 Airbus A321wing with spoilers visible, image courtesy of Charlotte Voisey

required to achieve this compared to producing the same maximum strength in one orientation only. It is far better to design a laminate with a ply lay up sequence chosen to match loading conditions.

6.3.2.3 The Materials Science of Fibre Reinforced Composites

Composite properties are affected by which material is used for the reinforcement, how much of it is present, how it is distributed and what length of fibre is used. All of these are fixed when the material is consolidated into a component, and cannot be changed.

Fibre Material

Whilst carbon fibres have the most impressive properties, they are not the only option. Other materials regularly used in FRP include aramid fibres, such as Kevlar®, and glass fibres. Table 6.1 shows key properties of the most important materials used for reinforcement in FRP, with mild steel also included for comparison. It must be noted that these values refer to the stand alone reinforcements, not the composite material. The relative costs should be regarded as indicative values only. The high cost of CFRP is the main reason why this material does not dominate the sector.

It is clear that carbon fibre stands out as having superior specific stiffness, though aramid fibres actually have the highest specific tensile strength. The much lower cost glass fibres have good properties for their price.

Not all carbon fibres are the same. The precise properties will vary depending on the details of the manufacturing technique, with the Young's modulus varying

Table 6.1 Typical properties of reinforcement materials for FRP, axial values shown for strength and stiffness

Material	Density ($g\ cm^{-3}$)	UTS (MPa)	Young modulus (GPa)	Specific tensile strength ($GPa\ g^{-1}\ cm^3$)	Specific stiffness ($GPa\ g^{-1}\ cm^3$)	Relative cost
Carbon fibre	1.86	3000–7000	230–500	1452	161	120
Aramid fibres (Kevlar 49)	1.45	2900	130	2000	89.7	60
E-glass fibre	2.54	2200	70	866	27.6	6
Mild steel	7.8	650	207	83	26.5	1

between 230 and 500 GPa. Higher quality carbon fibres are more fully graphitised. Fibre stiffness increases with graphitisation temperature, but so too does cost. Strength however peaks at a graphitisation temperature of approximately 1600 °C.

Glass is a notoriously brittle material, any existing flaws will easily propagate as cracks, leading to fracture. Glass fibres however are famously flexible, with widely circulated images showing knots tied in them. This is only possible when great care is taken in the handling of the glass fibres (Chap. 7). They are still very sensitive to flaws, but these are largely eliminated by applying a protective coating, known as size, immediately after drawing of the fibres. This protects them from mechanical damage such as scratches, and also from any chemical interaction with the atmosphere.

Fibre Fraction

The amount of reinforcement present impacts the overall properties of the composite, with composite properties improving as fibre fraction increases. The fibre fraction is the volume fraction of fibres within the composite. The impact of fibre fraction on overall composite stiffness, E_{comp} can be quantified via the rule of mixtures, where v_f is the volume fraction of fibres and E_f and E_m are the Young's moduli of the fibre and matrix material respectively.

$$E_{comp} = v_f E_f + (1 - v_f) E_m \tag{6.1}$$

The equation as written applies to long, aligned, fibres and to stiffness measured in the direction of alignment. If the fibres are orientated randomly then there will be uniform properties, which can be an advantage, but, for the same volume fraction of fibres, the stiffness will be lower than the maximum possible for aligned fibres. For random fibre distributions the rule of mixtures is modified by an efficiency factor, η_o, inserted before the $v_f E_f$ term. For 2D random fibre orientation a value of 3/8 is used, whereas for 3D random fibre orientation it is 1/5.

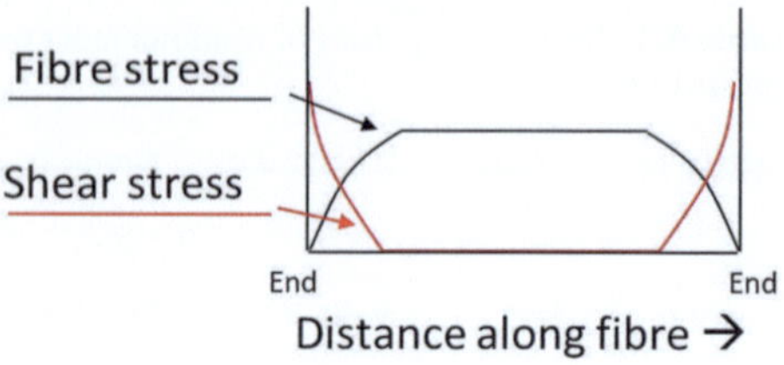

Fig. 6.18 Schematic showing how the stresses in fibres are modified at the fibre ends, decreasing the average stress and hence load carried by the fibres

Fibre Length

Fibre length also impacts overall composite properties. That is because the stress transfer from matrix to composite is less effective at the end of the fibres. The load carried in the ends of the composite is less than average (Fig. 6.18). Shorter fibre composites have more fibre ends per unit volume, hence end effects are more important. The shorter the fibres are, the larger the proportion of material that is effected by end effects. Long fibre composites have superior mechanical properties compared to shorter fibre composites because end effects have less impact in long fibre composites.

Whilst long fibres are good for mechanical properties, they restrict which processing techniques can be used. Resin infiltration is more challenging in long fibre composites and long fibre composites cannot be injection moulded, short fibre composites can be injection moulded but only when present in very low volume fractions.

Matrix Material

The polymer matrix transmits the applied stress to the reinforcement. It also has additional roles such as protecting the fibre surfaces and holding the reinforcement in place. A low viscosity matrix is advantageous regarding ease of infiltration. The matrix also acts as a crack propagation barrier, improving the toughness of the material. The matrix also determines the maximum service temperature as this is significantly lower than that of the reinforcement.

The matrix is chemically inert once completely cured, but during curing there is a variation of properties with time. The widely used epoxy matrix has an exothermic cure. The heat given out by the reaction limits component thickness. If the temperature increases to 180 °C or above then thermal degradation of the matrix can occur.

Matrix-Fibre Interface

The strength of the matrix-fibre interface affects the failure of the composite. This interfacial strength can be controlled by coatings. A coupling agent can increase the interfacial strength, which will in turn increase the transmission of stress to the fibres, improving the overall composite strength. However, a stronger interface is not always desirable. Other coatings can be used to deliberately weaken the interface. Doing this can increase the overall toughness of the composite by enabling the fibre pullout mechanism to operate.

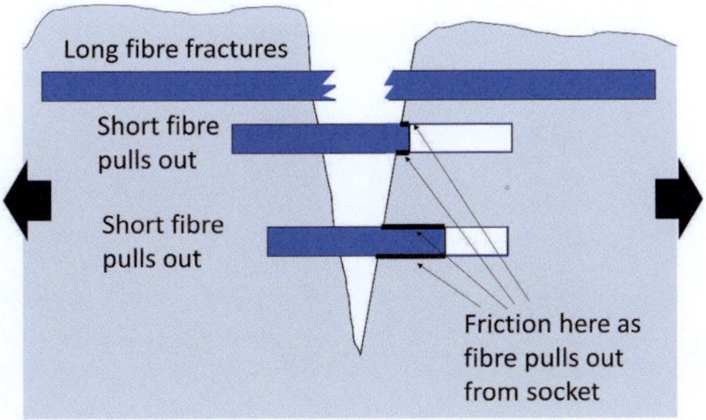

Fig. 6.19 Schematic diagram illustrating the fibre pull out mechanism which results in increased toughness due to the frictional forces that need to be overcome as fibres pull out from sockets

Fibre Pull-Out Mechanism

As a crack moves through the composite and hits a fibre, it is deviated around the fibre, debonding it from the matrix. This increase in crack path length results in a slight increase in the toughness of the material, the amount of energy required for a crack to grow. A more significant increase in toughness can be achieved via the fibre pull out effect. This occurs when, with the growth of a crack through the material, the crack opens, pulling fibres out of the sockets in the matrix where they have debonded (Fig. 6.19). The energy required to overcome the frictional forces generated as the debonded fibre ends slide out of the matrix represents a significant increase in the material toughness. For this mechanism to work the fibres need to debond and slide out of the sockets, rather than fracturing. The frictional forces increase with fibre length. This means that in long fibres the force required for fracture may be less than that to overcome the frictional forces. In this case the fibres fracture instead of sliding, and there is no increase in toughness. The fibre pull-out mechanism is therefore associated with short, not long, fibre composites. There is a critical length L_c below which the stress in the fibres cannot reach the fracture strength due to end effects and L_c changes with applied stress.

The significant increase in toughness resulting from the fibre pull out mechanism is exploited in energy absorbing applications of fibre composites such as in automotive crumple zones and aerospace fan casings [20].

Composite Architecture

The anisotropic, directional, properties of fibre composites have already been highlighted. Strength and stiffness are maximised when unidirectional fibres are used, but the properties are highly directional, with transverse properties being significantly worse than those measured parallel to fibre direction. Composite architecture is the arrangement of fibres within the matrix. This is frozen in during manufacture and cannot be subsequently changed.

In practice the properties can be somewhat evened out by using layers, or plies, with fibres aligned within a ply but with different plies being rotated with respect to each other. The overall distribution of fibre alignment directions will depend on the operating conditions. Fibre stacks have their own notation that shows the

Fig. 6.20 CFRP. Bottom half of images shows dry fibre fabric, the white threads hold the plies together. The upper ply is seen to have horizontal fibre tows, at the edge of the material the ends of vertical tows of underlying plies are visible. Consolidated CFRP is seen in the top half of the image, this has been formed from bidirectional pre-preg, the two perpendicular directions of fibre tows are visible within the surface

variation of fibre direction. Tailoring the stacking sequence is one way of generating a composite architecture to match operational requirements. It is also possible to have bidirectional layers where fibre tows (tows are yarns consisting of thousands of individual fibres) are woven, the most simple have sets of fibres at 0° and 90° (Fig. 6.20). The most complex are 3D weaves [21] but cost increases with complexity.

In addition to the different arrangements of fibres within the matrix (Fig. 6.20), fibre composites can be incorporated into other composites and sandwich structures, producing composite composites.

GLARE is laminate composite where layers of carbon prepreg are interleaved with layers of aluminium 2024. The overall material has a 20% weight saving compared to using the aluminium alloy by itself. This weight saving is attractive to the aerospace industry where GLARE has various applications including the upper fuselage skin of the Airbus 380 and the Boeing 777 bulk cargo floor.

Sandwich panels are another form in which composites are commonly used. The FRP faces of the sandwich panels are separated by a lightweight foam or honeycomb core which improves energy absorption whilst keeping overall density low. The sandwich panels act as extended I beams, with the high stiffness composite material increasing the overall stiffness of the composite panels. Sandwich panels have applications in Formula 1 and aerospace where their high stiffness, low density, good insulation and damping properties are exploited.

6.4 Test Your Understanding—Questions

Q1 Which polymers, thermoplastics or thermosets, can melt when heated?
Select one:

a. thermoplastics
b. both thermoplastics and thermosets
c. thermosets

Q2 Which of the following describes the bonding in a thermoset?
Select one:

a. weak
b. no intermolecular bonding
c. permanent strong chemical bonds

Q3 Compared to thermoplastics, thermosets are typically which of the following?
Select one:

a. lower strength, softer and more ductile
b. easier to recycle
c. harder, stronger and more brittle

Q4 Give three automotive examples of the use of non-reinforced polymers. In each case explain, with reference to material properties and operating conditions, why this is a suitable material choice.

Q5 State and explain two potential disadvantages of increased use of carbon fibre composites in the domestic automotive industry.

Q6 For a fixed reinforcing fibre fraction state what would allow the following properties to be maximised.

a. toughness
b. stiffness

Q7 Explain how and why weakening the matrix/fibre interface can improve a polymer composite and state a transport application that could benefit from this improvement.

Q8 Give two aerospace applications of non-reinforced polymers. In each case state the key material property that makes the polymer suitable for the application.

Q9 To make a composite with a stiffness of 80 GPa what volume fraction of reinforcing phase would be needed for:

a. single walled carbon nanotube composite?
b. carbon fibre composite?

	Stiffness (GPa)
Single walled carbon nanotube	1250
Carbon fibre	300
Polyethylene	0.2

You should state any assumptions.

Q10 Design considerations require a polymeric water reservoir to be housed under a car bonnet near the engine. The reservoir has a complex geometry due to needing to fit around other components. Two candidate polymers could be used, one is a thermoplastic, one is a thermoset. With reference to operating conditions and material properties state which should be used.

Q11 A thin, moulded automotive polymer panel is found to be distorted after being used at 150 °C. Explain why the distortion could have occurred and how the manufacturing process could be modified to prevent this happening.

Q12 State an application of carbon fibre composite that requires a high toughness. Explain why a high toughness is required.

6.5 Test Your Understanding—Answers

Q1 Answer a thermoplastics.

Q2 Answer c permanent strong chemical bonds.

Q3 Answer c harder, stronger and more brittle.

Q4 Answer Various alternative correct answers possible—example answers with expected level of detail given below:

Car lighting systems: a low load application that requires a transparent material, no high temperature requirements.

Car seating: low stiffness of polymeric foams, together with low density makes them suitable for seating applications.

Electrical insulation: low electrical conductivity of polymers coupled with the ease with which they can be shaped makes them suitable.

Q5 Answer Lack of ease of recyclability: EU legislation requires extensive reuse/recycling of car components, the intimate mix of materials in CFRP makes recycling complex and costly; Cost: inherent cost of CFRP and specialised manufacturing processes combined with the sensitivity of the domestic car market to cost is a disadvantage.

Q6 Answer

a. toughness: use of short fibres and a coating to weaken fibre/matrix interface to enable the fibre pull out mechanism to be exploited.
b. stiffness: unidirectional alignment of fibres in direction that stiffness needs to be maximised in.

Q7 Answer Weakening the matrix/fibre interface can increase the toughness of the polymer composite by enabling the fibre pullout mechanism to occur. The frictional forces between disbonded fibres and the surrounding socket in the matrix which must be overcome for crack opening to occur increase the energy required for crack propagation, hence toughness is increased.

Various correct answers regarding application including: F1 monocoques.

Q8 Answer Various correct answers including:

Aircraft windows—optical transparency.
Aircraft seating—low density.

Q9 Answer Assumptions: good load transfer across fibre/matrix interface—not currently easy for nanotube, also need to state if aligned or random fibres assumed (n.b. if random appropriate efficiency factor, n, needs to be included in rule of mixtures so that E_f is replaced by nE_f 2D random n = 3/8, 3D random n = 1/5).

Calculation below assumes aligned fibres

$$Ecomp = v_f(E_f - E_m) + E_m$$

$$\text{Need to find } v_f,\ v_f = (E_{comp} - E_m)/(E_f - E_m)$$

$$= (80 - 0.2)/(1250 - 0.2) = 6.4\% \text{ for nanotube}$$

$$= (80 - 0.2)/(300 - 0.2) = 26.6\% \text{ for carbon fibre}$$

Q10 Answer Operating conditions: temperature can be assumed to be moderate, and therefore potentially high enough to cause distortion if a thermoplastic is used – this is a disadvantage of the thermoplastic. The thermoset won't distort with exposure to temperature however manufacture will be more costly as blow moulding can't be used. Either answer accepted as correct as long as suitable supporting argument given.

Q11 Answer This is thermally induced distortion, the operating temperature of 150 °C provides thermal activation energy that enables molecular motion. Any molecules that have been frozen into a straight conformation during moulding will rearrange into a more convoluted molecular shape, shrinking dimensions and if anisotropic (likely) also distorting component shape. Diagram as below expected but not necessary if suitably clear written explanation.

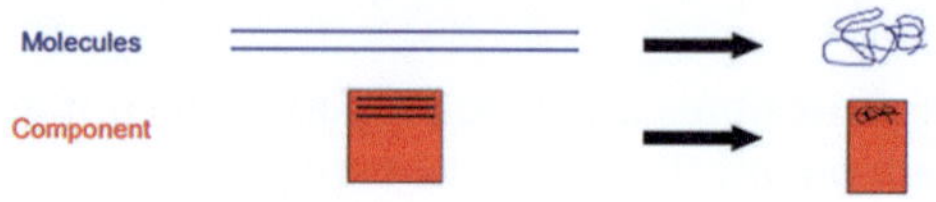

Modification of manufacturing process (either answer + reason acceptable); Allow longer in mould cooling times to decrease amount of frozen in extended molecules—therefore less distortion will occur—Use a thermoset instead of a thermoplastic—cross links will limit amount of molecular motion and hence distortion Not acceptable: decrease component usage temperature (would work but does not answer the question).

Q12 Answer Example answers showing expected level of detail are below, alternative correct answers are also possible.

Application: engine housing/fan case—high toughness required to absorb energy in a blade off event.

Or

Automotive side impact bars—high toughness required to absorb energy in the even of a crash.

Further Reading

Organisations and companies

Polymerminds.com aims to bring together the global polymer community. Its website links to many resources including research papers and has a comprehensive list of worldwide associations related to polymers https://www.polymerminds.com/polymer_associations.asp

Composites UK is the UK trade association for the fibre reinforced polymer sector, its website has many openly accessible resources relating to applications and processes as well as relevant events and sector strategy documents. https://compositesuk.co.uk/

SGL Carbon has useful information on manufacture of carbon fibres and CFRP applications on their website https://www.sglcarbon.com/en/carbon-fibers-and-cfrp/

Open access articles and research papers

The plastics section of the American Chemistry Council's website has many articles on current issues relevant to polymer materials. https://www.americanchemistry.com/better-policy-regulation/plastics

A Review of Fibre Reinforced Polymer Structures, Jawed Qureshi, Fibers 2022, 10(3), 27; https://doi.org/10.3390/fib10030027

UK Composites Industry Competitiveness and Opportunities, Lucintel, Dec 3 2020, https://iuk.ktn-uk.org/wp-content/uploads/2021/07/Opportunities-in-the-UK-Composites-Industry-Lucintel-Public-Version.pdf

Composites for electric vehicles and automotive sector: A review, Adil Wazeer, Apurba Das, Chamil Abeykoon, Arijit Sinha, Amit Karmakar, Green Energy and Intelligent Transportation, Volume 2, Issue 1, February 2023, 100,043, https://doi.org/10.1016/j.geits.2022.100043.

References

1. Yang Y, Jiang Y, Liang H, Yin X, Huang Y. Study on Tensile Properties of CFRP Plates under Elevated Temperature Exposure. Materials (Basel). 2019 Jun 21;12(12):1995. https://doi.org/10.3390/ma12121995. PMID: 31234381; PMCID: PMC6631531. https://www.ncbi.nlm.nih.gov/pmc/articles/PMC6631531/#:~:text=Through%20using%20new%20resins%2C%20special,(%E2%88%92196%20%C2%B0C).
2. https://www.evian.com/en_gb/our-sustainability-actions/packaging-and-recycling/ accessed 20/11/23
3. https://www.unilever.com/planet-and-society/waste-free-world/rethinking-plastic-packaging/ accessed 20/11/23
4. Regulations: end-of-life vehicles (ELVs), Guidance for manufacturers and importers. Office for Product Safety and Standards and Department for Environment, Food & Rural Affairs Published 1 April 2015, Last updated 1 January 2021 https://www.gov.uk/guidance/elv
5. https://www.bbc.co.uk/news/science-environment-61739159
6. https://www.ciel.org/issue/fossil-fuels-plastic/#:~:text=Just%20as%20the%20world%20begins,plastic%20industries%20are%20deeply%20connected accessed 27/11/23
7. Siracusa V, Blanco I. Bio-Polyethylene (Bio-PE), Bio-Polypropylene (Bio-PP) and Bio-Poly(ethylene terephthalate) (Bio-PET): Recent Developments in Bio-Based Polymers Analogous to Petroleum-Derived Ones for Packaging and Engineering Applications. Polymers (Basel). 2020 Jul 23;12(8):1641. https://doi.org/10.3390/polym12081641. PMID: 32718011; PMCID: PMC7465145.
8. A Brief Review of Technology and Materials for Aerostat Application, Bharathi Dasaradhan, Biswa Ranjan Das Biswa Ranjan Das's LiveDNA, Mukesh Kumar Sinha, Kamal Kumar, Brij Kishore and Namburi Eswara Prasad, Asian Journal of Textile, Year: 2018 | Volume: 8 | Issue: 1 | Page No.: 1–12 https://doi.org/10.3923/ajt.2018.1.12 https://scialert.net/fulltext/?doi=ajt.2018.1.12
9. https://www.hybridairvehicles.com/news-and-media/overview/news/airlander-10-s-hull-will-be-supplied-by-ilc-dover/
10. Have The Boeing 787 And Airbus A350 Set Standard For Composites? Thierry Dubois June 18, 2020 aviationweek.com https://aviationweek.com/aerospace/manufacturing-supply-chain/have-boeing-787-airbus-a350-set-standard-composites
11. https://flywith.virginatlantic.com/eu/en/stories/the-airbus-a350-a-quiet-efficient-giant.html#:~:text=Aerodynamics,different%20phases%20of%20the%20journey accessed 22/10/23

12. https://www.azom.com/article.aspx?ArticleID=8194
13. Composites World Automotive Published 12/22/2021 "BMW rolls out multi-material Carbon Cage with 2022 iX vehicle line" Hannah Mason https://www.compositesworld.com/articles/bmw-rolls-out-multi-material-carbon-cage-with-2022-ix-vehicle-line
14. Rolls-Royce starts manufacture of world's largest fan blades, made with composites, for UltraFan demonstrator, Ginger Gardiner, 2/11/2020, Composites World https://www.compositesworld.com/news/rolls-royce-starts-manufacture-of-worlds-largest-fan-blades-made-with-composites-for-ultrafan-demonstrator
15. Saeedi, A, Motavalli, M, Shahverdi, M. Recent advancements in the applications of fiber-reinforced polymer structures in railway industry—A review. Polym Compos. 2023; 1–21. https://doi.org/10.1002/pc.27817
16. Marine Application of Fiber Reinforced Composites: A Review, Felice Rubino, Antonio Nisticò, Fausto Tucci, Pierpaolo Carlone, J. Mar. Sci. Eng. 2020, 8(1), 26; https://doi.org/10.3390/jmse8010026
17. https://www.hybridairvehicles.com/our-aircraft/our-technology/
18. https://www.azom.com/article.aspx?ArticleID=21371
19. High-rate, automated aerospace RTM line delivers next-gen spoilers, Jeff Sloan, 7/1/2020, CompositesWorld. https://www.compositesworld.com/articles/high-rate-automated-aerospace-rtm-line-delivers-next-gen-spoilers
20. Fromm, Joshua. "Composite Fan Blades and Enclosures for Modern Commercial Turbo Fan Engines." (2016). https://www.colorado.edu/faculty/kantha/sites/default/files/attached-files/fromm.pdf
21. El-Dessouky HM and Saleh MN (2018) 3D Woven Composites: From Weaving to Manufacturing. Recent Developments in the Field of Carbon Fibers. InTech. Available at: https://doi.org/10.5772/intechopen.74311.

7 Brittle Materials

7.1 Introduction

Ceramics, concrete and glass are materials grouped together by a shared characteristic that is almost always a disadvantage, that of brittleness (Fig. 7.1). This makes them very sensitive to flaws and to poor performance in tension. These brittle materials have high strength in compression, but cannot withstand tensile stresses. However, with appropriate materials engineering the issue of brittleness can be effectively combatted, allowing these materials to be exploited in useful applications.

The applications and related characteristics of glass, ceramics and concrete are considered in turn. The end of the chapter explains the underlying materials science behind how engineering ceramics, ceramic matrix composite and glass fibres function as useful engineering materials.

7.2 Glass

Glass: The bullet points

- Chemically inert and optically transparent
- Glass can corrode, with even small amounts of corrosion impacting performance in optical components
- Lamination and residual stresses can be used to limit impact of brittle failure
- Elimination of defects allows the inherent high strength of glass to be exploited in glass fibres

K. T. Voisey, *The Engineer's Guide to Materials*,
https://doi.org/10.1007/978-3-031-62937-2_7

Fig. 7.1 A fractured ceramic mug. The same, fractured, mug is shown in each image. Left: Fragment has been removed to clearly show the crack path. Right: The fragment has been put back in place to illustrate the lack of deformation. This is how brittle materials usually behave: when they fracture all that happens is a crack propagates, nothing else happens, no change of shape, no deformation, no yield. This enables the fragment to perfectly fit back into place. and also means there is little energy barrier to crack propagation, i.e. the material is brittle, due to the lack of energy absorbing mechanisms, such as yield

Strictly speaking, glass is a state of matter rather than a particular material. If metals are cooled extremely quickly then metallic glasses can be formed. This section is using the term glass as it is generally used, to refer to silica based glasses. The transparency and chemical inertness of such glass is exploited in numerous applications. It is generally able to withstand continual exposure to the weather without requiring maintenance, making it a popular and widespread material for windows. By far, the vast majority of windows contain glass and there is extensive architectural use of glass, with glass forming the majority of the exterior of most large modern buildings. Automotive windscreens are made from toughened glass. The transparency of glass is also exploited in many optical applications such as lenses used in optical microscopes, binoculars and telescopes as well as enhancing the vision of many people by its use in spectacles. The low cost of glass, combined with its chemical inertness, has lead to many packaging applications. Engineering applications of glass include bridges made from structural glass and numerous applications which exploit the strength of glass by using it as the reinforcement in glass fibre reinforced polymer composites, commonly referred to as fibreglass.

7.2.1 Sustainability Considerations Relating to Glass

Glass is made from sand, a natural material which is available in abundant supplies. High temperatures, and the associated high energy input, is required in glass manufacture and has associated carbon emissions. The chemical inertness of glass means that glass objects can be used for very long periods of time, there are many glass windows that are centuries old. Such long lifetimes improve the sustainability of glass. The chemical inertness also means that any glass pollution will not result in any chemical environmental hazard, however there may clearly be physical hazards from sharp glass shards. Glass also operates at the other extreme, as a single use packaging material. Glass can be repeatedly recycled, making it highly sustainable and a number of European countries are recycling 95% of their waste glass [1]. A variety of different figures are reported for the UK's glass recycling rates, but they are all clearly someway below 95%. Issues with recycling glass include colour separation and the hazards associated with handling sharp glass fragments. The damage that such fragments can inflict on any co-collected material such as paper and cardboard can decrease the value of these other materials. Glass collection via bottle banks solves these problems which is why some UK councils collect glass in this way instead of by roadside collection.

7.2.2 Applications of Glass

Probably the most important current application of glass is something most people use everyday without being conscious they are making use of glass. This is the 5 billion kilometres of silica glass fibre optical communications cabling which supports the internet. This application makes use of the transparency of glass as well as its chemical inertness which results in long fibre lifetimes.

The flexibility of glass fibres is very different to how glass usually behaves (Fig. 7.2). The brittleness of glass means that it easily fractures if there are any defects present. Eliminating defects eliminates the problem. This is generally not viable for any macroscopic pieces of glass, however it can be achieved for glass fibres. Firstly, the length of any defects is restricted to the diameter of glass fibres. The small diameter of glass fibres automatically decreases the dimensions of the flaws that can be present. Ensuring that glass fibres have a protective coating, referred to as "size", applied as soon as they are drawn from the melt eliminates surface defects. The lack of defects mean that glass fibres can carry load without fracturing, enabling the high strength to be exploited. This leads to the use of glass fibres as a reinforcing material in glass fibre polymer composites which can be a very cost effective alternative to carbon fibre reinforced composites. Glass fibres are also highly flexible, again this results from the elimination of defects enabling the material to withstand the stresses generated as it is deformed without fracturing. This, along with the transparent nature of glass, leads to it use in communication where the flexibility of optical fibres enables easy construction of optical fibre networks.

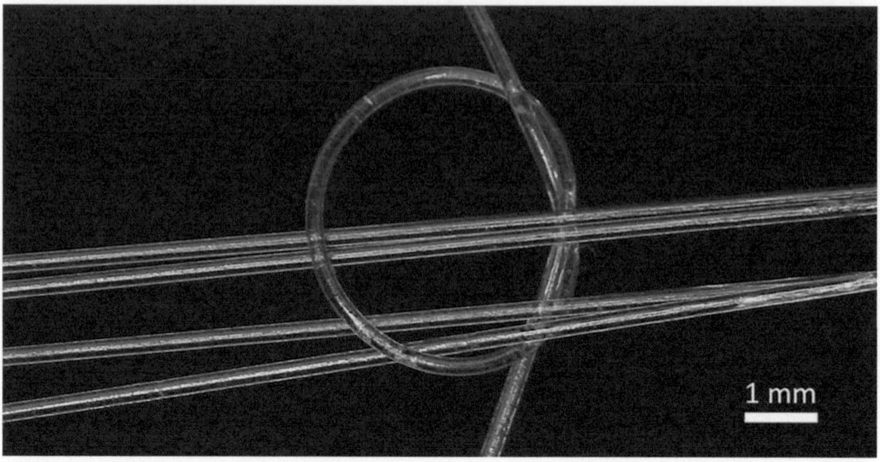

Fig. 7.2 Commercial silica glass optical fibre, 125 μm diameter fibre with protective polymer coating, giving a total diameter of 250 μm. The increase in flexibility seen is due to the increase in tensile strength which results from the elimination of defects. Image courtesy of David Furniss

The most familiar engineering application of glass is in window applications. These clearly exploit the transparency of glass but also rely on the chemical inertness of glass which results in good corrosion resistance and good weather resistance. Glass windows are continuously exposed to the weather, any surface chemical reaction could have significant impact on optical transparency. This becomes particularly important when glass is used in complex architectural structures where replacement is logistically challenging and potentially highly costly. Very small amounts of corrosion, which would be negligible if loss of load bearing material was the concern, can cause noticeable surface marks sufficient to require replacement of glass units.

With advances in glass manufacturing, ever larger glass windows are possible. The business districts of most cities feature glass clad high rise buildings. The glass needs to withstand wind loading as well as maintain transparency. Any failure needs to be contained, having sharp glass shards falling from height could have horrific consequences. Advances in structural glass makes this possible. Structural glass is glass with load bearing capacity. Structural glass can improve the amount of light, by maximising the amount of glass and minimising the support. This allows architects to better integrate structures with the landscape and to enhance connectivity between different areas of a building. It is also widely used in stand alone showpieces such as the numerous glass bridges around the world which act as tourist attractions. Structural glass is load bearing as it uses units of laminated glass.

Laminated glass is often used as a glazing option that is more secure. This uses a polymer interlayer between two layers of glass. This improves impact resistance. Even if the glass fractures, it will remain adhered to the polymer layer. This

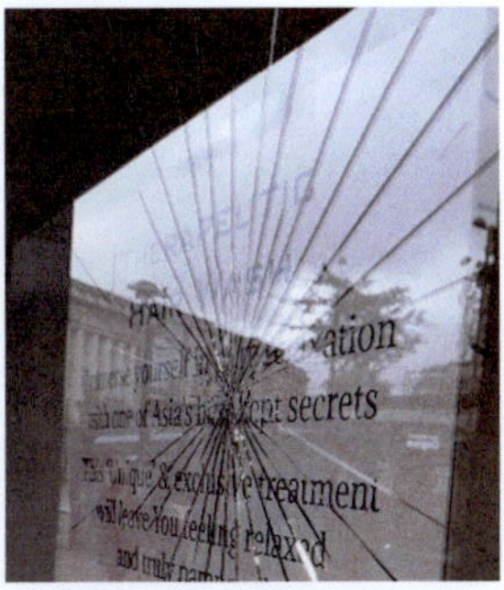

Fig. 7.3 Laminated glass used in a shop window. The retention of transparency after fracture is evident, as is the retention of material

eliminates the possibility of any large detached shards of glass, improving safety. This results in applications in overhead skylights, where there would otherwise be serious safety concerns from falling shards. It also enhances security as it is more difficult to form a complete breach (Fig. 7.3). Areas at risk from break in and burglary such as banks and jewellers frequently use laminated glass. Laminated safety glass, LSG, is used extensively in automotive windscreens and is a legal requirement in most countries. It is preferred to toughened glass due to the greater retention of structural integrity, which provides some additional passenger protection. LSG also retains greater transparency in the event of failure, compared to the finely crazed and essentially opaque appearance that toughened glass takes on after failure (Fig. 7.4). This gives the driver a better chance of manoeuvring their vehicle to safety in the event of windscreen failure.

Toughened glass is widely used in doors, architectural panels and bus windows (not windscreens). Residual stresses are exploited in toughened glass. During manufacture compressive stresses are generated on the outer surfaces, with balancing tensile stresses in the interior of the material. The compressive stresses are useful in making the surfaces more resistant to small cracks and defects, toughening the material. The tensile stresses are what produce the distinctive shattered appearance of failed toughened glass (Fig. 7.4). The tensile stresses ensure that any cracks penetrating the material propagate rapidly and easily, breaking the material up into small fragments which are safer than the large sharp shards that are generated by fracture of non-toughened plate glass.

A related application is in screens for electronic products. The transparency is again exploited, along with the high hardness, scratch resistance and corrosion resistance. Fracture resistance is improved by lamination or residual stresses as with window glass. Chemical toughening can also be used, and is used in Gorilla® Glass, this diffuses potassium ions into the surface region to generate compressive stresses, which then inhibit crack growth.

One application that exploits the brittleness of glass is in break glass fire alarm call points (Fig. 7.5) and the related application of emergency door open switches. Here glass is used as a material that will fracture easily. A glass panel suppresses a button, when the glass is fractured the button is released, activating the fire alarm

Fig. 7.4 A, failed, toughened glass side window in a bus. The fragmentation of the failed window, and the resultant crazed appearance, is evident

or releasing an electromagnetic door lock. To ensure reliable, and safe, fracture the glass panel has a built in pre-crack which both ensures fracture will occur under a reasonable applied force and guides the crack path to avoid generation of sharp glass shards.

Fig. 7.5 One of the applications of glass that exploits its brittleness, in a fire alarm break glass panel

7.3 Ceramics

Ceramics: The bullet points

- Chemically inert inorganic non-metallic materials
- Inherently brittle but toughness can be significantly increased by microstructural manipulation
- High melting points
- Low density

Ceramics are a group of materials that have a wide range of applications. Familiarity with their use in domestic crockery, which exploits their general chemical inertness, low thermal conductivity and good wear resistance, can make their more advanced applications surprising.

There are many uses of ceramic materials in sensors, electronics and solar panels. Ceramic dental implants are widely used. Other medical applications include use of ceramic materials for the load bearing surfaces of replacement joints. Hydoxyapatite ceramics have a number of bone tissue engineering applications.

Ceramic coatings are used in all gas turbine engines to protect metallic components along the hot gas path. The high hardness and wear resistance are exploited in cutting tools as well as in ceramic particles used as reinforcement in wear resistant coatings.

Engineering ceramics and ceramic/ceramic composites optimise the internal structure of the material in order to minimise the determinantal impact of the inherent brittleness of ceramics. This opens up industrial applications such as turbine blades and valve seats where properties such as high strength, wear resistance and high operating temperatures can be exploited.

7.3.1 Sustainability Considerations Relating to Ceramics

The majority of ceramics are chemically inert. This means though they may be present in the environment for a very long time there are no environmental concerns related to leaching of toxic substances. The majority of sustainability concerns related to ceramics are secondary effects resulting from the use of ceramics. Sensor applications of ceramics allow various different systems to run better, increasing their sustainability. The use of ceramic coatings in gas turbine engines permits higher temperature, more efficient, engine operation.

7.3.2 Applications of Ceramics

The high melting point and chemical inertness of oxide ceramics makes them well suited to high temperature applications which exploit the refractory nature

Fig. 7.6 Refractory bricks lining a rotary kiln to insulate the underlying steel shell from the high temperatures of the kiln."File:Refractory bricks lining.jpg" by Alexknight12 is licensed under CC BY-SA 3.0

of these materials. Refractory ceramic bricks are widely used to line furnaces and kilns (Fig. 7.6). The low thermal conductivity, typical of ceramics, minimises heat loss, saving energy.

Ceramics form the main layer of thermal barrier coatings, which are key to enabling nickel based superalloys survive the ever higher operating temperatures encountered in gas turbine engines (Fig. 7.7). The characteristic low thermal conductivity and high melting points of ceramics are the key properties here. The ceramic layer insulates the underlying metal from the high temperatures of the hot gas path. In a thermal barrier coating there can be a 200 K temperature drop across a 200 μm ceramic layer. The most widely used ceramic for this application is partially stabilised zirconia, selected due to its stability in the relevant temperature range and for its relatively high, at least for a ceramic, coefficient of thermal expansion. This minimises stress generated due to the mismatch of coefficient of thermal expansion between the ceramic coating and underlying metallic component. Gas turbine engine components protected by thermal barrier coatings include turbine blades, nozzles, nozzle guide vanes and combustion chambers. Related automotive applications are turbocharger casings and exhausts. They are also used to coat aluminium to make more effective heat shields.

High end automotive brakes in sports cars and supercars use carbon/ceramic composites materials due to their retention of performance at high temperature,

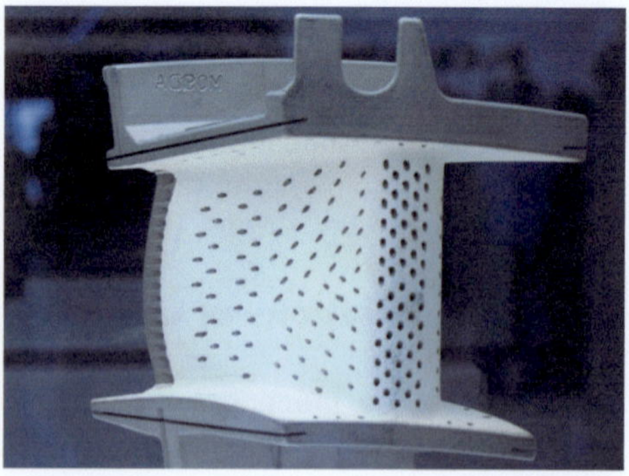

Fig. 7.7 The white ceramic thermal barrier coating is seen on the exterior of this V2500 turbine guide vane. Effusion cooling holes are also visible. "File:Repair process for a V2500 high-pressure turbine guide vane (10).jpg" by Olivier Cleynen is licensed under CC BY-SA 3.0

making them resistant to brake fade (carbon can be considered a ceramic). The material consists of carbon fibres embedded in silicon carbide. Porsche's ceramic composite brakes (Fig. 7.8) permit a 50% weight saving compared to steel alternatives, typically removing 20 kg from the unsprung mass of their sports cars [2]. The Lamborghini Urus, which in 2019 had the largest brakes ever used on a production car, used carbon/ceramic brake discs to enable braking from 100 kph to a full stop in under 35 m [3]. Formula 1 [4] uses carbon/carbon brake discs due to the more demanding operating conditions and greater requirements for heat dissipation. Here carbon fibres are embedded in carbon, forming a composite that is significantly less brittle than monolithic carbon. The interfaces within the material act to stop easy crack propagation, hence toughening the material making it less brittle than monolithic carbon. With toughness improved, this allows the inherent high strength and wear resistance of ceramics to be exploited in brake applications. Carbon/carbon brakes are also used widely in aviation, having been used widely in commercial aircraft since the 1980's, including on the Boeing 747, 767, 777 and 787 Dreamliner [5]. Key advantages of both carbon/ceramic and carbon/carbon brakes are their good durability, resulting from their high wear resistance, and their low mass compared to conventional cast iron and steel based systems. Use of carbon, instead of steel, brakes for the Boeing 747 resulted in a weight saving of 820 kg [6].

Fig. 7.8 Porshce's distinctive yellow calliper ceramic composite brakes use SiC reinforced with carbon fibre. "File:Porsche Ceramic Composite Brake with silicon carbide—Museum fur Naturkunde, Berlin—DSC09917.JPG" by Daderot is marked with CC0 1.0

Ceramic matrix composites, CMCs, have long been a target material to replace nickel based superalloys in gas turbine engines. CMCs have the high temperature capability to cope with the hot gas path, and to allow even hotter engine temperatures. Their lower density, approximately a third that of nickel based superalloys, means that valuable weight savings are possible. A lot of work has gone into improving these materials so that they can cope with the very high stresses generated in rotating engine parts. The superior properties of CMCs compared to normal, monolithic, ceramics is due to their interior structure (Figs. 7.9 and 7.10). The multiple interfaces that this finescale microstructure contains restricts flaw size and inhibits crack propagation. The aerospace industry has focussed on SiC/SiC (Fig. 7.10) with Rolls-Royce investing in developing this material for its Ultrafan engine [7] and General Electric using SiC/SiC CMC in the shrouds in the hot section of their LEAP engine [8, 9]. There is further use of this material in the GE9X engine which uses SiC/SiC CMC in five different parts including combustors and nozzle applications. These are again components that need to withstand the aggressive environment of the hot gas path hence environmental barrier coatings are required to protect the underlying material [9].

There are many uses of ceramics in machining tools where again the combination of high hardness and wear resistance makes them attractive. Cermets are used

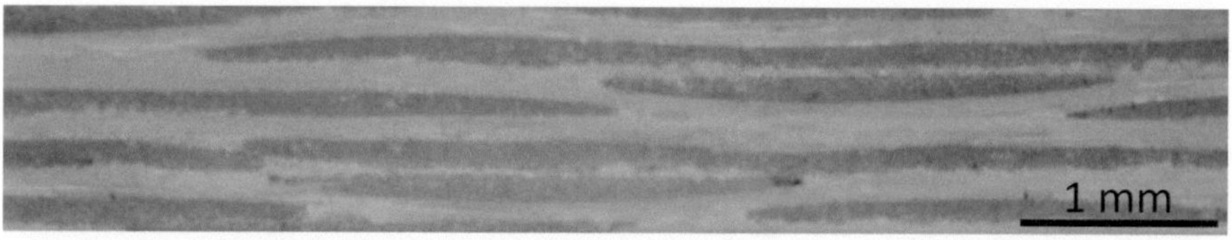

Fig. 7.9 Optical micrograph of a cross-section of an alumina/alumina (ox/ox) ceramic composite. The woven structure is apparent, the darks areas are tows, threads consisting of hundreds of individual alumina fibres. Image courtesy of The Manufacturing Technology Centre

Fig. 7.10 Optical micrographs of a cross-section of a SiC/SiC silicon carbide ceramic composite. The woven structure is apparent. The higher magnification image on the right allows individual fibres within the tows to be seen. Images courtesy of The Manufacturing Technology Centre

Fig. 7.11 The use of cermet give this cutting blade a hard, wear resistant, cutting edge. "File:Metal-cutting-blade.JPG" by User:Billbeee is licensed under CC BY 2.5

widely for such applications. Cermets are composites where a hard, wear resistant ceramic reinforcement is embedded in a metallic material. Cermets are an example of metal matrix composites. The cermet material is tougher, less brittle, than the ceramic as though any cracks may extend easily through the ceramic particles they will terminate when they reach the interface with the tougher metallic constituent of the material. Tungsten carbide, WC, and titanium carbo-nitride, TiCN, are popular ceramics in cermets, with WC cermets being particularly well suited for cutting tools (Fig. 7.11) and TiCN cermets for machining [10].

Catalytic convertors need large surface areas for efficient operation. They also have demanding, chemically aggressive, high temperature operating conditions. The high melting points and chemical inertness of oxide ceramics make them well suited to cope with these conditions. Ceramics have found applications as the substrate material which supports small particles of the active precious metals. Cordierite (a magnesium-alumino-silicate ceramic) is widely used, it is extruded into monolithic substrate structures containing many fine internal channels, with diameters less than 5 μm [11].

Ceramics have numerous applications in sensors. Their underlying properties make ceramics particularly well suited for this. There is extensive use of sensors in the automotive industry. Even low end models such as the Ford Fiesta feature about twenty different types of sensors including parking aids, oxygen sensors and air bag impact sensors. The lambda sensor is used to measure oxygen levels in automotive exhausts in order to control the air/fuel mix and hence decrease emissions. Zirconia and titania are two oxide ceramics widely used in lambda sensors. These sensors are based on electrochemical principles, specifically the Nernst equation. This can be used to determine how much an electrode potential will change as a function of the chemical environment. As the oxygen level changes the magnitude of the voltage generated by the ceramic oxide electrode changes, and measurement of this allows the oxygen level to be known [12]. Capacitive and piezoelectric properties of ceramics are exploited in various pressure sensors and actuators. Piezoelectric materials have been used in fuel injection systems to open the injectors in a highly controlled way. They are also widely used in parking detectors. Here vibrations of a piezoelectric generates the ultrasound wave which is used to determine the distance of any nearby obstacles [13]. A similar application has been used for the fuel additive tank on various diesel vehicles, here the piezoelectrically generated ultrasound wave is used to determine the level in the tank [13].

Alumina is widely used as a substrate in the electronics industry. In this application the electrically insulating nature of alumina is exploited.

7.4 Cement and Concrete

Cement and concrete: The bullet points

- Chemical reaction → changing properties with time
- Size and form of aggregate impacts overall properties
- Concrete generates a significant amount of heat during setting
- Steel reinforcement widely used to enhance tensile strength

Concrete is the most widely used material, with particular usage in civil engineering applications. 30 billion tonnes are used globally each year [14]. The popularity of concrete is due to the combination of low cost and attractive properties including durability, fire resistance, compressive strength and impact resistance. As with other brittle materials, performance in tension is a concern however inclusion of steel within reinforced concrete, particularly in the locations which will experience tensile stresses, improves this.

Concrete is a composite material, being formed of cement, water and aggregate. A key feature of concrete is that it forms by chemical reaction between its constituent materials. The heat released and the progressive change in properties during setting are factors that need to be taken into consideration during design and planning of concrete construction. It is standard practice to wait for 28 days after pouring before measuring the strength of concrete, it will usually reach about 70% of this strength within seven days though the hardening process can continue for years. Both of these factors need to be taken into account, particularly during the construction of large structures.

The overall properties of concrete can be tailored in a number of ways including modifying the aggregate, using additives (also referred to as admixtures), changing the cement used as well as changing the relative proportions of the constituents. The versatility of the material is evident in the range of applications, more unusual applications including decorative concrete bowls (Fig. 7.12) and the entries to the

a

b

Fig. 7.12 **a**, **b** Decortative concrete items, images show work from @kathyshawconcrete and are included courtesy of Kathy Shaw

annual American Society of Civil Engineers National Concrete Canoe Competition (Fig. 7.13). The majority of usage is in large infrastructure projects.

7.4.1 Sustainability Considerations Relating to Concrete and Cement

There are significant current concerns regarding the sustainability of concrete, due to the large associated energy input. 8% of global CO_2 emissions are attributed to concrete manufacture [15]. The high temperatures required during cement making have a high energy input. A large proportion of concrete's CO_2 emissions are due

Fig. 7.13 Concrete canoes. **a** Cal Poly's winning canoe from 2019 in action, **b** Detail of Cal Poly's winning canoe from 2023, where the concrete surface texture is evident. Images courtesy of Cal Poly's concrete canoe team

to fuel burn associated with the heating of cement kilns. However, the major contributor to concrete related CO_2 emissions is the direct release of CO_2 during the calcination process, part of the cement manufacturing process. There is considerable work being done in the area of low carbon concrete including using industrial by-products such as steel mill slag as a substitute for cement. Concrete can be recycled, or rather downcycled. Crushed concrete from demolition sites can be repurposed as aggregate for new construction, with extensive use in road building and airfields [16]. Use of concrete can also be positively exploited in the design of energy efficient buildings [17] where its large thermal mass can contribute to decreased energy use for heating and cooling.

7.4.2 Applications of Concrete and Cement

Dams are some of the largest applications of concrete. China's Three Gorges Dam (Fig. 7.14), operational from 2012, is the largest hydroelectric project in the world and used almost 30 million cubic metres of concrete. The high compressive strength, low cost, good weather resistance and long lifetime are the characteristics of concrete that made it a suitable choice here.

There is extensive use of reinforced concrete in residential and commercial building construction (Fig. 7.15). There is also extensive usage of this form of concrete in large infrastructure projects such as coastal defences and structures, bridges and roads. Properties of reinforced concrete relevant to these applications are its fire resistance, durability and thermal and acoustic insulation properties, as well as the generally low cost. An important advantage of this composite material compared to non-reinforced concrete is that the high tensile strength of the steel reinforcing bars compensates for the poor performance of concrete in tension.

Fig. 7.14 China's Three Gorges Dam, a large civil engineering application of concrete. "Three Gorges dam" by hughrocks is licensed under CC BY-SA 2.0

Fig. 7.15 "Steel reinforcement prepped for a concrete pour for the future LIRR Concourse floor. (CH014B, 9-12-2017)" by MTA C&D—EAST SIDE ACCESS is licensed under CC BY 2.0

7.4.3 How Concrete Works

Concrete does not simply dry out. Water actively participates in chemical reactions with cement to form concrete. The overall reaction is summarised below.

$$\text{Cement} + \text{water} \rightarrow \text{concrete} + \text{heat}.$$

Depending on the precise chemistry of the cement there are numerous different specific reactions which contribute to the hardening of concrete. The chemical reactions occurring as concrete sets are exothermic: they give out energy as heat. The heat released during the chemical reaction cannot be ignored. In large scale structures such as dams the heat released needs to be taken into account with thermal management plans.

Due to the chemical reactions, concrete is much more than simply a physical mixture of components. The inner structure of the material undergoes significant changes, which, if done correctly, form the desired interconnected structure. These reactions do not occur instantaneously, they take time. The structure and properties of concrete change significantly while the reactions are happening.

Cement is not just one single chemical compound, and different cements have different chemical mixes. The key constituents are tricalcium silicate and dicalcium silicate. These both react with water. In the first stage the silicates dissolve in water and form an ion rich gel. Solid phases then progressively form, precipitating out from the gel, allowing further ions to dissolve. As the solid phases grow they form a solid, interconnected, three dimensional network which holds the aggregate in place (Fig. 7.16).

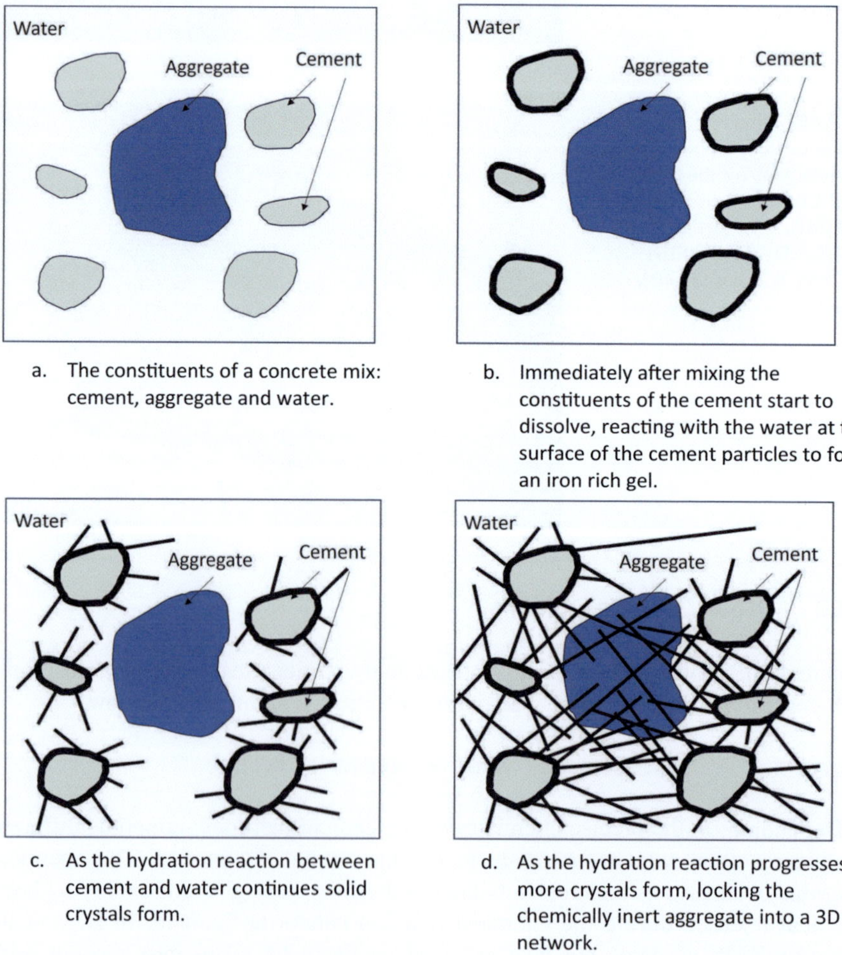

a. The constituents of a concrete mix: cement, aggregate and water.

b. Immediately after mixing the constituents of the cement start to dissolve, reacting with the water at the surface of the cement particles to form an iron rich gel.

c. As the hydration reaction between cement and water continues solid crystals form.

d. As the hydration reaction progresses more crystals form, locking the chemically inert aggregate into a 3D network.

Fig. 7.16 Schematic diagrams showing evolution of concrete structure during hardening

The cement, aggregate and water content of the mix needs to be appropriate in order to produce good quality concrete. There needs to be sufficient water present for the reactions with cement to occur. In practice, more water than this minimum is required to ensure that the concrete mix has good workability and so can be moved into moulds easily. However, if the water content is too high a lower strength concrete will be produced.

As with any composite material there are challenges relating to dispersion of reinforcement. Increasing the proportion of aggregate should improve the mechanical properties. However, this can also make proper mixing more challenging. To result in a properly bonded concrete, each piece of aggregate should be coated in cement. During setting water chemically reacts with the cement, binding the aggregate into the material. This requires good mixing of the water, cement and aggregate. If the cement does not come into contact with water it will remain unreacted and will not be consolidated into the overall material. The aggregate does not directly participate in the chemical reactions. It becomes bound in place by the rigid structure that results from the reaction of water and cement.

The time taken for the chemical reactions to complete means that the material properties progressively change until the chemical reaction is completed. This evolution with time needs to be taken into account when planning, particularly relating to the time before the concrete can be regarded as having load bearing capacity. Concrete is usually poured into forms, temporary moulds, which are removed once the concrete is set (Fig. 7.17). Good understanding of the setting process is required in order to know when it is appropriate to remove the formwork.

A variety of additives, or admixtures, can be used to modify concrete. Water-reducing admixtures decrease the required water content, whilst maintaining the high workability required to enable the liquid concrete to flow, producing a higher strength concrete without the cost of increasing cement content. Retarding admixtures extend the time required for setting, This can be useful in increasing the amount of time that concrete is workable for, including the time available to transport concrete to site. Conversely, accelerating admixtures decrease the setting time which can be useful in cold climate work where the lower temperatures slow the chemical reactions. Current areas of research include the use of novel additives including nanomaterials, fibre reinforcements and waste plastic.

Concrete properties are also influenced by the aggregate. The shape and size of aggregate has a strong influence on the flow characteristics of concrete. This is an important consideration as flow of concrete is required during construction. Smaller, spherical aggregate is better for flow and workability. Having a range of sizes of aggregates increases the overall packing efficiency, which decreases cost due to lowering the proportion of cement in the concrete. Large aggregates require less water but there can be issues with large aggregates settling out of the mix. Angular aggregates, such as those obtained by crushing stone, have a larger surface area which can enhance the bonding of the aggregate into the overall material. The aggregate surface area also impacts the setting time of the concrete.

Porosity, and water, are always present in concrete. In some cases higher porosity levels are deliberately induced in order to decrease the density of the concrete.

Fig. 7.17 Wooden formwork will need to be kept in place until the concrete setting reactions have proceeded enough for the concrete to be self supporting. "Formwork and rebar in advance of a concrete pour at the East Approach for the planned Amtrak Westbound Bypass Tunnel which will send trains below the existing LIRR tracks. (CH057A, 9-13-2017)" by MTA C&D—EAST SIDE ACCESS is licensed under CC BY 2.0

One example of this is Reinforce Autoclaved Aerated Concrete (RAAC). This hit the UK headlines in 2023 [18] where there were concerns about the material failing. RAAC can be used as an effective building material, however, as with all materials, it needs to be used appropriately. In the case of RAAC the issue was insufficient sealing against water ingress. This caused problems as it accelerated the corrosion of steel reinforcing bars. The greater porosity means that there is less material for water and aggressive species to diffuse through before reaching the steel. Aggressive species therefore reach the steel more quickly, initiating corrosion earlier and hence decreasing material lifetime.

Sacrificial anodes and impressed current systems can be used to protect the steel reinforcement from corrosion. For this to work there needs to be electrical connection to the reinforcement, which needs to be set up before the concrete is poured. Once the steel reinforcement has started to corrode it is difficult to repair. Additional, externally applied reinforcement may be required.

7.4.3.1 The Materials Engineering of Brittle Materials

Materials engineering strategies that enable effective use of brittle materials are based on either of two principles: (1) Limitation of crack growth. (2) Elimination of defects. It is successful implementation of such strategies that enables glass fibres, engineering ceramics and ceramic matrix composites to be used in engineering applications.

The many applications such as machining tools and tarmac that make use of the hardness and wear resistance of ceramic and stone reinforcement respectively are successful as they embed particles of the brittle material within another, tougher, material. The overall result is a composite material. The material, within which the brittle reinforcement is located, does more than simply hold the reinforcement in place. It is really important that the brittle reinforcement material is present as discontinuous, discrete, particles. Cracks may grow through the particles but are stopped when they reach the edge of the particle. The interface with the surrounding material acts as a barrier to further crack growth. This limitation of crack growth to the particle size prevents the growth of defects into large cracks which would occur in a monolithic block of the brittle material. The importance of interfaces in making crack growth more difficult is highlighted in ceramic/ceramic composites where, despite each of the underlying materials being inherently brittle, a usefully crack resistant composite can be produced. This can even be true when the reinforcement and surrounding material are the same, such as in the carbon/carbon composites used in Formula 1 brake systems. In terms of materials properties, the fracture toughness of the material has been increased: the energy required to propagate a crack through the material has increased.

This same crack limiting effect frequently occurs in pavements, where cracks run through individual paving slabs but terminate at the edges (Fig. 7.18). It is simply a larger scale example of the effect.

Fig. 7.18 Cracks in stone pavement slab terminating at edge of slab

Fracture strength is one of a set of different strengths: yield strength, ultimate tensile strength, shear strength, tensile strength, compressive strength... All of these refer to the maximum stress that a material can withstand before something happens. For fracture strength this is the maximum stress a material can withstand before fracture by crack propagation. Fracture strength is not strictly a material property as it varies depending on what defects are present. The larger the size of any flaws are, the lower the fracture strength is, and vice versa.

A simplified version of the derivation of the fracture strength equation is given below, the Griffith criterion. This is included to show why fracture strength and flaw size are related.

This derivation considers the two key things that happens when a crack grows. These are that new crack surface is created and that some material is unloaded. The generation of new crack surface requires energy whereas the unloading of previously loaded material releases stored elastic strain energy. The fundamental requirement that must be met for a crack to grow is simply that the strain energy released has to be sufficient to form the new crack surface.

This statement will be converted into equations. The system being considered, is a material of thickness t subject to a tensile stress σ. An internal crack of length 2a which passes through the whole thickness of the material and which is oriented perpendicularly to the applied stress is going to grow by a small amount, δa, at each end (Fig. 7.19). It is assumed that a circular area of material around the crack is unloaded due to the presence of the crack. Noting that the crack extends throughout the thickness, t, of the material means that the unloaded material actually forms a cylinder, of radius a, extending through the material around the crack. When the crack grows the radius of the cylinder increases by δa.

The new crack surface generated as the crack in Fig. 7.19 grows by δa at each end is given by the expression below (7.1), where the new crack length is 2δa,

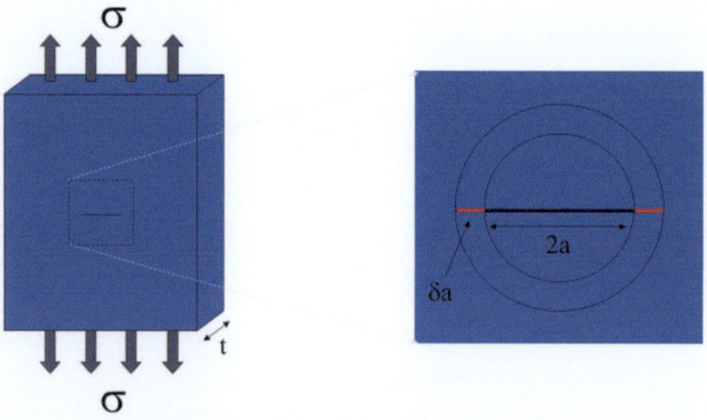

Fig. 7.19 Crack considered in the derivation of the fracture strength equation

this runs through the whole thickness of the material hence is multiplied by material thickness t to give the new crack surface area. It is then really important to remember that the new crack has two surfaces, the upper and lower surfaces, the second 2 in the expression takes this into account. The final symbol, γ, is the energy required per unit area of new crack surface, multiplying by this gives the total energy required to form the new crack surface for the situation shown in Fig. 7.19:

$$\text{Energy required} = 2\partial a t 2\gamma \tag{7.1}$$

Determination of how much energy is released as the crack grows requires consideration of the stored elastic strain energy per unit volume. This is equal to half the stress, σ, multiplied by the strain, ε, giving $\frac{1}{2}\sigma\varepsilon$. The relevant volume is the area between the two concentric circles shown in Fig. 7.19, multiplied by the thickness of the material, t. A good approximation to that area, which works when ∂a is small compared to a, is simply to multiply the circumference of the inner circle, $2\pi a$, by ∂a. Multiplying this additional unloaded volume by the stored strain energy per unit volume gives the energy released by the crack growth considered in Fig. 7.19:

$$\text{Energy released} = \frac{1}{2}\sigma\varepsilon 2\pi a\partial a t = \frac{1}{2}\frac{\sigma^2}{E}2\pi a\partial a t \tag{7.2}$$

For the crack to grow the energy released has to be at least enough to provide the energy required to generate the corresponding new crack surface. Using the expressions derived so far, this requirement can be stated as:

$$\frac{1}{2}\frac{\sigma^2}{E}2\pi a\partial a t \geq 2\partial a t 2\gamma \tag{7.3}$$

which simplifies to

$$\frac{\sigma^2}{2E}\pi a \geq 2\gamma \tag{7.4}$$

If the rough approximation of a circular unloaded region around the crack is replaced by more accurate consideration of the stress field, the factor of 2 by which the left hand side of the expression is divided disappears. The resultant expression can be made more general by replacing 2γ, which only takes surface energy into account, by G_c, which is a broader term including any other energy absorbing mechanisms that accompany fracture. For brittle materials such as glasses and ceramics, $2\gamma = G_c$, whereas for materials where there are additional energy absorbing mechanisms such as fibre pull out in composites or dislocation motion in metals, $2\gamma < G_c$. G_c is the toughness and has units of J m^{-2}.

Implementing these changes and rearranging for the stress, σ, gives the expression below. This states that the applied stress, σ, has to exceed the expression on the right in order for crack propagation, i.e. fracture, to occur.

$$\sigma \geq \sqrt{\frac{G_c E}{\pi a}} \tag{7.5}$$

The minimum stress for which this is true is the fracture strength, σ_f, the maximum stress the material can withstand before fracture:

$$\sigma_f = \sqrt{\frac{G_c E}{\pi a}} \tag{7.6}$$

This expression shows how the fracture strength, σ_f, varies with flaw size, a. The impact of decreasing flaw size on increasing the fracture strength, σ_f, can be quantified.

It can be useful to further rearrange this expression as below. This separates the variables and material properties, producing a new materials property, K_c, the fracture toughness, or critical stress intensity factor, which has units of MPa $m^{1/2}$. Whilst fracture strength varies with flaw size, the fracture toughness, K_c, remains constant. The derivation is based on a single flaw of known size. An unstated assumption was that the flaw was perfectly sharp, Y is a dimensionless geometry factor used to take flaw geometry into account. In general, materials will have a flaw size distribution, not a single flaw. It is appropriate for a statistical, probability based, approach to be used in related calculations. The method used is Weibull statistics.

$$\sigma_f Y \sqrt{\pi a} = \sqrt{G_c E} = K_c \tag{7.7}$$

Materials engineering actively makes use of this relationship between flaw size and fracture strength.

In fibre glass the glass fibres have small fibre diameters which restrict the dimensions of what flaws can be present, increasing the fracture strength in line with Eq. 7.6. In addition to this, great care is taken to ensure that the surfaces of the glass fibres are protected from damage by coating them immediately after they are drawn from the melt. As fibres flex, the greatest stresses are generated in the surface regions. Fibres with flaw free surfaces can be highly flexible, as, in the absence of flaws, the inherently high strength of glass can be exploited.

Engineering ceramics have higher fracture strengths due to minimisation of flaw size. Engineering grade silicon carbide, SiC, is chemically identical to, but more expensive than normal SiC. What is different, and is responsible for the superior material properties, as well as the higher price, is the internal microstructure of the material.

Ceramics are made by sintering of powder. In sintering the powder particles are heated up and stick together to form a rigid part by diffusion which links together adjacent particles. Broadly speaking, the length scale of typical defects in the sintered structure will be approximately the same as the size of the starting powder. Engineering ceramics simply start with a finer starting powder, leading to smaller defects and hence a higher fracture strength. The underlying principle is straightforward. However, it needs to be noted that a finer powder adds complication and cost. The material needs to be broken up or ground down into smaller particles, which increases cost. There can also be increased handling difficulty for finer particles.

Ceramic matrix composites, CMCs, combine two materials engineering strategies to enhance their properties. As with glass fibres and engineering ceramics, CMCs have a structure which minimises the dimensions of flaws and thereby increases the fracture strength. CMCs also benefit from the fibre pull out mechanism, (see Sect. 6.3.2.3). The additional energy required to overcome the frictional forces as fibres pull out of sockets during crack opening increases the fracture toughness of the material.

7.5 Test Your Understanding—Questions

Q1 With reference to material properties and operating conditions, explain why engineering ceramics meet the material requirements for automotive engine valves.

Q2 A silicon carbide automotive engine valve must withstand a stress of 300 MPa. The fracture toughness, for silicon carbide is 4 MPa $m^{1/2}$. Two different forms of silicon carbide, SiC, are available: i, refractory grade, maximum flaw size = 120 μm; ii, engineering grade, maximum flaw size = 12 μm; For each grade determine the fracture strength and state if the grade would be suitable for use in the engine valve. Assume the geometry factor, Y, = 1.

Q3 Cermet tipped cutting tools are widely used. What is a cermet and why are these materials well suited to this application?

Q4 What enables optical glass fibres to bend rather than snap?

Q5 State two applications of glass fibre, indicating the key properties in each case.

Q6 Explain how gas turbine aero engines rely on ceramics.

Q7 Briefly state the three main constituents of concrete and what their roles are.

Q8 Give an example where materials engineering enhances the toughness of a brittle material by limiting crack growth.

Q9 How do engineering ceramics differ from normal ceramics?

7.6 Test Your Understanding—Answers

Q1 Answer Operating conditions are reasonably high temperatures approx. 350 °C, a corrosive environment and cyclic stresses. A corrosion resistant material with good fatigue resistance, and or wear resistance, that can operate at 350 °C is therefore required. Engineering ceramics can withstand the operating temperatures and are hard and wear resistant so can cope with the harsh surface conditions.

Q2 Answer

$$K = \sigma\sqrt{(\pi a)}$$

Refractory grade: $\sigma_f = 4/\sqrt{(\pi \times 120 \times 10^{-6})} = 206$ MPa cannot be used as valve.
Engineering grade: $\sigma_f = 4/\sqrt{(\pi \times 12 \times 10^{-6})} = 651$ MPa can be used as valve.

Q3 Answer A cermet is a cemented carbide, this consists of ceramic particles embedded in a metal. For the cutting tool application, the high hardness of the ceramic particle reinforcement make a durable cutting tool. Any cracks stop at the edge of the particles, not travelling through the more ductile metal, making the cermet far tougher than the monolithic ceramic.

Q4 Answer The glass fibres are defect free, their surfaces are protected by a coating preventing any surface defects, the fracture strength is increased so that the underlying high strength of glass can now be exploited, enabling significant bending stressed to be induced without fracture.

Q5 Answer Various different answers possible, two examples given:

Optical fibres—here the optical transparency is required, along with the high strength which enables optical cables to have the flexibility to be routed.

Reinforcement in fibre glass—here it is the high strength of glass, which is available to be exploited due to the defect free nature of the glass fibres.

Q6 Answer Ceramic thermal barrier coatings are required to protect the metallic components along the hot gas path, without these engine operating temperatures would be significantly lower, resulting in less efficient engines and component lifetimes would be significantly shorter, adding to engine maintenance costs.

Q7 Answer Cement—reacts with water to form a solid network holding the aggregate in place.

Water—chemically reacts with cement to bond aggregate in place.

Aggregate—acts as a strengthening reinforcement.

Q8 Answer Various possible correct answers, most likely one given:

Ceramic/ceramic composites—here the many interfaces within the material act as crack terminators, preventing easy crack growth beyond the interface.

Q9 Answer

The key difference is the microstructure, no difference in composition. Engineering ceramics simply have smaller flaws, which results in a larger fracture strength.

Further Reading

Glass Technology Services have information about applications of glass and various other resources available via their website https://www.glass-ts.com/

The British Ceramic Federation has many ceramic related resources https://www.ceramfed.co.uk/

https://great-ceramic.com/ has a lot of data relating to their own ceramic products as well as more general resources on various aspects of advanced ceramics.

Advanced ceramic components: Materials, fabrication, and applications, Oti-toju, TA, Okoye, PU, Chen, GT, Li, Y, Okoye, MO, Li, SX, Journal of Industrial and Engineering Chemistry, Volume 85 Page34-65 2020 https://doi.org/10.1016/j.jiec.2020.02.002 – open access review paper on advanced ceramics.

The American Concrete Institute has various resources, particularly in the "topics in concrete" section of the website.

There is a useful introduction to concrete accessible via Linkdin: Cement & Concrete Chemistry Primer, Mike Stanzel, Jan 2 2019 https://www.linkedin.com/pulse/cement-concrete-chemistry-primer-mike-stanzel/

References

1. https://gaskellswaste.co.uk/recycling/glass-reprocessor/ accessed 27/11/23
2. Porsche Ceramic Composite Brake (PCCB) May 30, 2022 / By Richard Lindhorst elferspot.com https://www.elferspot.com/en/magazin/porsche-ceramic-composite-brake-pccb/
3. Lamborghini Urus has the World's Biggest Brakes, Ron Cline, 27 January 2019, 6th gear Automotive Solutions. https://www.6thgearautomotive.com/2019/01/27/lamborghini-urus-has-the-worlds-biggest-brakes/#:~:text=And%20the%20Lamborghini%20Urus%20delivers,a%20thickness%20of%2040%20millimeter.
4. Formula 1 Innovations: Brabham's Carbon Brake Revolution, Lawrence Butcher, May12, 2020, Motorsport https://www.motorsportmagazine.com/
5. Operational Advantages of Carbon Brakes, Tim Allen, Trent Miller, Evan Preston, Operational Advantages of Carbon Brakes aero quarterly qtr_03 | 09, Aero magazine, published by Boeing Commercial Airplanes https://code7700.com/pdfs/operational_advantages_carbon_brakes.pdf
6. https://en.wikipedia.org/wiki/Boeing_747-400 accessed 27/11/23

7. https://www.rolls-royce.com/media/press-releases/2018/16-07-2018-rr-reaches-future-technology-milestone-as-government-confirms-further-funding.aspx accessed 27/11/23
8. https://www.geaerospace.com/future-of-flight/landmark-technologies accessed 27/11/23
9. Ceramic matrix composites taking flight at GE Aviation, Jim Steibel, American Ceramic Society Bulletin, Vol. 98, No. 3, (April 2019) p30–33 https://ceramics.org/wp-content/uploads/2019/03/April-2019_Feature.pdf
10. Cermet Systems: Synthesis, Properties, and Applications, Subin Antony Jose, Merbin John and Pradeep L. Menezes, Ceramics 2022, 5(2), 210–236; https://doi.org/10.3390/ceramics5020018
11. https://dieselnet.com/tech/cat_subs_cer.php accessed 27/11/23
12. https://autoditex.com/page/lambda-sensor-o2-sensor-21-1.html accessed 27/11/23
13. The Piezo Electric effect – David Wagstaff AAE MIMI Master Technician, By Autotech-Nath, June 7, 2018 https://autotechnician.co.uk/the-piezo-electric-effect-david-wagstaff-aae-mimi-master-technician/
14. Concrete needs to lose its colossal carbon footprint Nature Editorial 28 September 2021 https://www.nature.com/articles/d41586-021-02612-5
15. Concrete: 8% of global emissions and rising. Which innovations can achieve net zero by 2050? January 24, 2023 by Ben Skinner and Radhika Lalit, energypost.eu. https://energypost.eu/concrete-8-of-global-emissions-and-rising-which-innovations-can-achieve-net-zero-by2050/#:~:text=Concrete%20manufacture%20is%20responsible%20for,mainly%20driven%20by%20developing%20nations
16. https://www.concretecentre.com/Performance-Sustainability/Circular-economy/End-of-life-recycling.aspx accessed 27/11/23
17. Concrete for energy efficient buildings The benefits of thermal mass, European Concrete Platform ASBL, April 2007. https://www.theconcreteinitiative.eu/images/ECP_Documents/ConcreteForEnergyEfficientBuildings_EN.pdf
18. Reinforced Autoclaved Aerated Concrete (RAAC): frequently asked questions, first published 8 sept 2023, last updated 19 oct 2023. https://www.gov.wales/reinforced-autoclaved-aerated-concrete-raac-frequently-asked-questions

Usage Induced Changes to Materials in Service

8

Usage induced changes: the bullet points

- Material properties and dimensions can change in service, with it being possible for multiple mechanisms to be active at any given time
- It is not only metals that undergo materials degradation
- Important to be aware of localised corrosion when considering the impact of corrosion rates
- Humid air can transport water to unexpected places
- Electrochemical corrosion is driven by potential differences, these can arise due to non-obvious variations in conditions
- Metal component with fluctuating stresses → need to consider fatigue
- Metal component under load at high temperature → need to consider creep
- Analytical tools exist to help predict wear, corrosion, fatigue and creep lives

8.1 Introduction

There are many different mechanisms of materials degradation and corrosion. Published comments on costs relating to corrosion report numbers that are typically 2–3% of GDP, with world costs now somewhere in the region of $2.5 trillion [1]. The majority of materials degradation is due to the various mechanisms of degradation that metallic materials can undergo, however it is not only metallic materials that can degrade in service, all materials can.

Various different in service phenomena can result in changes to the material. These changes impact material properties and hence overall performance. Such changes can be due to chemical interactions with the environment, exposure to high temperatures or other aspects of conditions encountered in service. These

K. T. Voisey, *The Engineer's Guide to Materials*,
https://doi.org/10.1007/978-3-031-62937-2_8

changes are almost always detrimental and undesirable. The impact that any such changes have on the future performance of components needs to be understood. The different degradation mechanisms are discussed separately but it must be noted that several mechanisms may be active simultaneously in service, resulting in accelerated degradation.

Clearly, changes in operating conditions can generate problems, particularly if loads or temperatures are outside of planned ranges. What is focussed on here are changes resulting from exposure to the expected operating conditions. These are generally progressive changes, but can result in sudden failure.

Degradation of materials due to interaction with the environment can result in material loss, chemical changes and physical changes. The resulting changes in properties and dimensions can impact the performance of the components or structures, ultimately leading to failure. Corrosion, and related materials degradation phenomena, is an important consideration but is not the only one. Situations in which fatigue, creep and wear may occur, and their impact on performance, also need to be understood.

The benefit of understanding the causes and mechanisms of material degradation is to enable appropriate material selection and materials management techniques and strategies to be implemented. The result is optimised material usage. Steel is a highly important engineering material despite being notoriously susceptible to corrosion. The fact that steel is so widely used is proof that there are effective corrosion management strategies. An overview of general corrosion management is given in Chap. 9.

The key point made in this chapter is that materials can change in service and that associated checking is required to ensure on going safe performance, Chap. 9 includes information on various non-destructive testing methods that can be used for such checking.

8.2 Sustainability Considerations Relating to Usage Induced Changes Materials in Service

Understanding the material degradation mechanisms highlighted in this chapter enables better material usage. By ensuring that materials are well suited to the operating conditions, material wastage associated with premature failure is minimised, improving sustainability.

8.3 Metallic Corrosion

There are examples where corrosion occurs and is not a problem. There are many examples of structures with visible rust which continue to function without significant problems (Fig. 8.1). However, there are other cases where localised corrosion can lead to failure due to a small amount of material being concentrated in a single location: the amount of corrosion that will lead to failure of electronic devices is

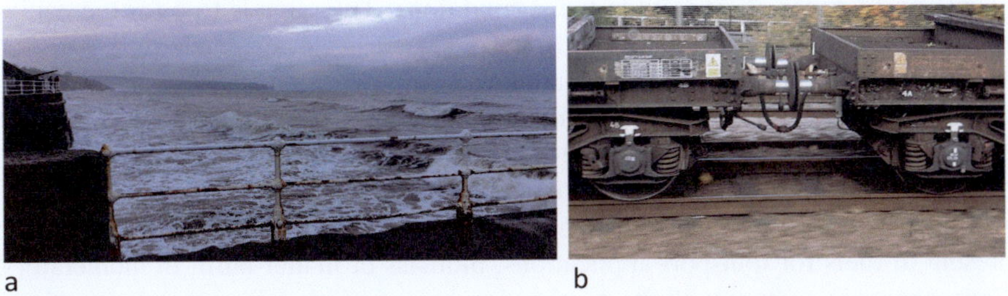

Fig. 8.1 Rust is clearly visible on the **a** rails and on **b** rails and steel parts of the rail bogie and freight car bases, however in both cases this level of rust is not considered problematic for these applications

negligible on a large structure such as a bridge. Similarly, localised corrosion in a pipeline can cause failure by leaking after a relatively small amount of material has been lost, far less than would be required for structural collapse of the same pipeline. This highlights the importance of understanding whether uniform or localised corrosion is active when determining the impact of reported or measured rates of corrosion.

Metallic corrosion can be split between high temperature oxidation and electrochemical corrosion. This can broadly be regarded as a division between dry, high temperature oxidation, and wet, electrochemical, corrosion. This is a convenient division when considering the different applications and environments in which metallic corrosion is a concern, however high temperature oxidation and electrochemical corrosion are both part of a continuum of behaviour. High temperature oxidation can be regarded as an electrochemical process.

For metallic corrosion, while there are various different corrosion mechanisms, the underlying issue is always that useful metal is being converted into detached metallic ions which no longer usefully contribute to the metallic structure or component. The progressive loss of material with time decreases load bearing capacity, can lead to leaks and changes in component dimensions all of which can lead to failure.

8.3.1 High Temperature Oxidation

Applications where high temperature oxidation needs to be considered are any conditions when a metal is heated in contact with air, or an oxygen containing environment.

The hot gas path through a gas turbine engine (Fig. 8.2) includes numerous metallic components which need to be resistant to high temperature oxidation. Gas temperature starts to rise in the compressor, from this point onwards high temperature oxidation is a concern. Compressor blades, combustors, turbine blades and nozzle guide vanes all operate in these conditions. The same is true for the equivalent components within land based turbines used in power generation. Fireside high temperature oxidation of boiler tubes is another application where this is a concern within power generation (Fig. 8.3a). Here the conditions are complicated by fly ash erosion and the presence of other aggressive species such as sulphur, particularly when waste or biomass is being used. Hot corrosion refers to specific corrosion mechanisms that are active in the presence of sulphur, which can be present in fuels for transport applications, biomass or matter burnt in incinerators. Components in contact with the combustion gases in internal combustion engines, the pistons, cylinder liners and exhaust system also need to be resistant to high temperature oxidation.

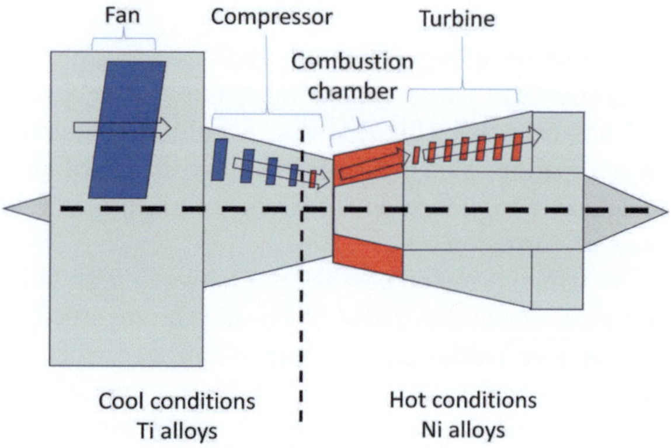

Fig. 8.2 Schematic diagram of a gas turbine engine, components needing high temperature oxidation resistance are those along the hot gas path, namely the hot end of the compressor, the combustors, and components in the turbine

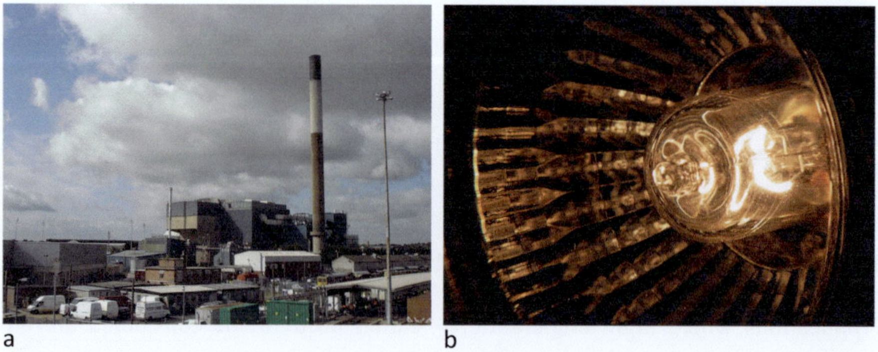

Fig. 8.3 Examples of where metallic components operate at high temperatures and the potential of high temperature oxidation has to be taken into account. **a** The Nottingham waste incinerator, the boiler tubes within are protected by coatings, **b** A light bulb filament is protected by being housed in an inert environment, **a** 948663 "File:Nottingham Waste Incinerator Facility—geograph.org.uk—948663.jpg" by Oxymoron is licensed under CC BY-SA 2.0. **b** "Light Bulb" by tolomea is licensed under CC BY 2.0

Metallic heating elements, and the filaments in incandescent light bulbs also need to be resistant to high temperature oxidation (Fig. 8.3b). Heating elements are generally made from inherently oxidation resistant material such as high chromium steel. That is an economically viable valid option for this application as the quantities of material used are not high. For light bulbs oxidation is prevented by eliminating oxygen from the environment by filling the bulbs with an inert gas such as argon. This method is very effective in this application but it is not possible to widely implement it.

In all these cases, the underlying issue is that oxygen reacts with the surface of the metal to form an oxide. If oxidation continues then there is progressive metal loss and high temperature oxidation becomes a problem.

In materials such as stainless steels this oxide formation is desirable as it protects the material from further oxidation.

8.3.2 Stainless Steels

Stainless steels work by reacting with oxygen in the environment to generate a surface oxide layer. In normal conditions all metals, except gold, do this. The special thing about stainless steels is that the oxide layer formed acts as a barrier to further oxidation. It does this by being adherent and defect free. This is characteristic of chromium oxide, chromia, forming on steel. Stainless steel is deliberately alloyed with high levels, greater than 11% of chromium, in order to ensure there is enough chromium present for the oxide layer to form. Overall, stainless steels undergo a small amount of oxidation in order to grow a barrier which prevents any further oxidation, by isolating the underlying steel from the environment (Fig. 8.4a). The amount of material consumed to produce the protective oxide is miniscule and has no detrimental effect on the load bearing capacity of any macroscopic component. The stainless steel is protected from corrosion as long as the protective oxide layer forms and remains continuous.

For the majority of other metals and non-stainless steels a surface oxide layer is also formed on exposure to the environment. The difference is that this is not protective. The oxide forms spalls off, exposing fresh metal to the environment, which then oxidises and spalls (Fig. 8.4b). This cyclic process continues with corresponding consumption of the underlying metal.

The high levels of chromium in stainless steel are required to ensure that the oxide layer is self healing if any damage occurs. The self healing oxide leads to stainless steel being widely used in applications such as heat exchangers, furnace components, fluidised beds, combustion chambers and burners. The oxide layer is also effective at protecting the underlying material from electrochemical corrosion, leading to extensive use of stainless steel in the chemical, petrochemical and food industries, particularly in piping, where aggressive chemical environments are encountered.

Chromium, aluminium and titanium, and their alloys have generally good corrosion and oxidation resistance as the oxides formed share these characteristics of

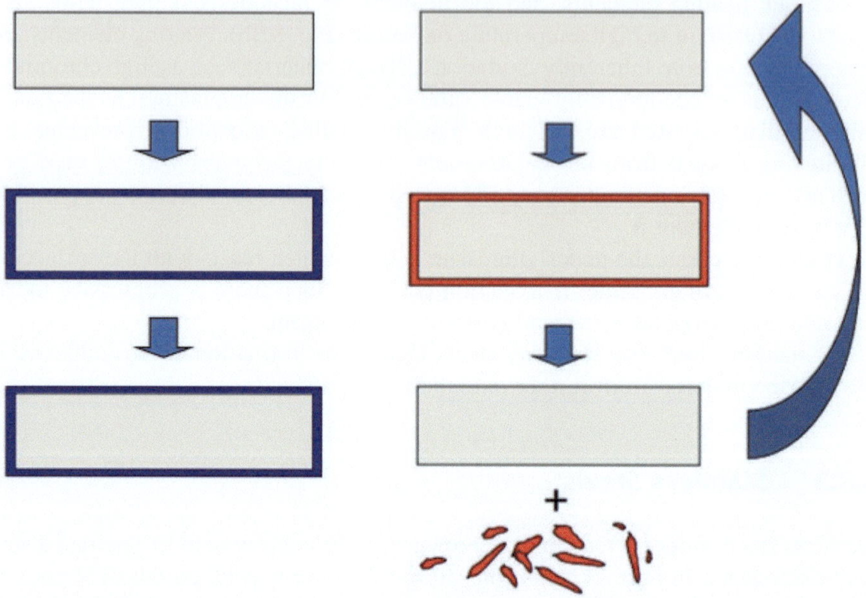

a. Stainless steel b. Non-stainless steels and other metals

Fig. 8.4 Schematic indication of **a** the growth and retention of a protective oxide layer on a stainless steel, **b** the repeating cycle, and associated material loss, of oxide formation, spallation and regrowth for non-stainless steels

being adherent and defect free and hence effective barriers against further degradation. These metals are frequently used to enhance corrosion resistance of alloys. This is why nickel based superalloys and coatings used to protect against high temperature oxidation have chromium and aluminium as alloying elements.

The likelihood of corrosion depends on the combination of material and the details of the environment. Stainless steels are known for their corrosion resistance, however there are conditions which stop stainless steels from being stainless.

If oxygen levels are too low then a protective oxide layer cannot be formed and the stainless steel is no longer stainless. This can occur in specific chemical environments as well as low oxygen conditions. If the temperature is too high then the oxide destabilises and is no longer protective. If the oxide is damaged in an aggressive environment then the underlying steel may start corroding before the oxide layer is repaired. Chloride ions and strong acids are particularly notorious for being able to damage the oxide layer. Situations which include either of these conditions can allow corrosion of the stainless steel to occur, it essentially becomes non-stainless. The Gibbs free energy and the Nernst equation are tools that can be used to understand the effect of conditions on oxidation and corrosion behaviour.

8.3.2.1 How Conditions Affect Oxidation and Corrosion—Gibbs Free Energy and the Nernst Equation

Gibbs free energy and the Nernst equation are closely related tools that are used to analyse any given environment to determine which way a chemical reaction will proceed, in the forwards or backwards direction. These equations can be used to determine at which temperature a metal will stop reacting with oxygen to form an oxide, and the oxide will instead decompose to reform the metal. This ability of reactions to change direction, for corrosion reactions to proceed or reverse, depending on conditions is true for all materials, and all chemical reactions. The Nernst equation can be used to determine the conditions under which gold, famously corrosion resistant, corrodes.

For Gibbs free energy, G, the requirement is that the reaction runs in the direction that produces a decrease in Gibbs free energy, i.e. ΔG is negative. In Eq. 8.1 ΔG is the change in Gibbs free energy, ΔG^0 is the value for standard conditions, R is the universal gas constant 8.31 J K^{-1} mol^{-1}, T is the temperature in Kelvin and K is the reaction quotient.

$$\Delta G = \Delta G^0 + RT \ln K \tag{8.1}$$

$$E = E^0 - \frac{RT}{zF} \ln K \tag{8.2}$$

The Nernst equation (Eq. 8.2) is the electrochemical version of this. However, as electrons, which are negatively charged, are involved there is a change in sign and the requirement is that the overall potential difference is positive for the forward reaction to occur. Here E is the electrode potential, E^0 is the standard electrode potential, R and T are again the universal gas constant and temperature in Kelvin respectively and K is again the reaction quotient. z is the number of electrons involved in the reaction and F is Faraday's constant, 9.6485×10^4 C mol^{-1}. The factor −zF can be used to convert between E and ΔG. Values of standard electrode potentials and changes in Gibbs free energy can be found in chemistry text books or easily accessible on line resources.

In both cases the effect of non-standard temperature and chemical environments are taken into account. Temperature, T, features clearly in both equations. The chemical environment is embedded within the reaction quotient K which takes into account concentrations of reactants and products. In some texts Q is used in place of K. For the general chemical reaction in Eq. (8.3), where a moles of A react with b moles of B to give c moles of C and d moles of D, the reaction quotient is given by Eq. (8.4), where square brackets indicate concentration: [C] is the concentration of C.

$$aA + bB \leftrightarrow cC + dD \tag{8.3}$$

$$K = \frac{[C]^c[D]^d}{[A]^a[B]^b} \tag{8.4}$$

The equations are included here to highlight that both temperature and chemical environment influence chemical reactions, and that these effects can be quantified.

It should also be noted that increasing temperature will generally increase corrosion rates, and hence decrease corrosion lifetime, as it accelerates the underlying chemical reactions. However, if the increase in temperature results in complete loss of the electrolyte due to evaporation then corrosion will be stopped as the electrochemical circuit is broken.

8.3.3 Electrochemical Corrosion

Broadly speaking, electrochemical corrosion is a concern when metals are in contact with water. It is important to understand that this is not only restricted to metal in direct contact with liquid water. The water content in air can also result in electrochemical corrosion. Humid air can transport water into unexpected locations, with subsequent condensation generated pooling of water in places not expected to be in direct contact with water. Applications where electrochemical corrosion of metals is a concern therefore include not only situations where metallic components are in contact with aqueous solutions but also where there is contact with air, due to the presence of air borne water vapour due to relative humidity.

Relevant applications are numerous and include metallic water piping, food and chemical industry applications as well as any metallic structures that are continuously exposed to the outside environment such as bridges, staircases, marine applications, shipping, oil rigs as well as metallic automotive, aerospace and rail components.

Regardless of the mechanism, the same underlying reaction is occurring in the electrochemical corrosion of a metal, M:

$$M \leftrightarrow M^{z+} + ze^- \tag{8.5}$$

In the metal dissolution reaction, Eq. (8.5), it can be seen that the rate of electron production is in direct proportion to the consumption of metal. There is a fixed ratio, z, between the number of metal ions lost from the component and the number of electrons released. The rate of electrons produced is therefore in proportion to the corrosion rate. Measuring the number of electrons per second is simply a measure of electrical current. This is why corrosion currents are measured, they are a direct measure of corrosion rate.

The two headed arrow in Eq. (8.5) indicates that the reaction can occur in either direction. The direction that the reaction runs in depends on the temperature and chemical environment. The direction of reaction can be determined using the Nernst equation (Eq. 8.2). For electrochemical corrosion to occur Eq. (8.5) needs to run in the forward, left to right, anodic direction. This releases electrons. A cathodic, electron consuming reaction is also needed.

One such reaction would be the reverse reaction of Eq. (8.5), though to get overall net corrosion a different reaction is required. The equivalent reverse version of Eq. (8.5) for a different metal often acts as the cathodic reaction. In galvanic corrosion, also referred to as dissimilar metal corrosion, or bimetallic corrosion this is what happens. It is the mechanism that underlies the operation of batteries. The reduction of dissolved oxygen, Eq. (8.6), is the most common electron consuming reaction that features in electrochemical corrosion. Oxygen and water are reactants, this is why oxygen and water are so closely associated with corrosion.

$$O_2 + 4e^- + 2H_2O \leftrightarrow 4OH^- \tag{8.6}$$

For electrochemical corrosion to occur an electrically connected anode and cathode are needed, as well as an electrolyte which allows the completion of the electrochemical circuit. The current is carried by ions in the electrolyte and electrons in the metal (Fig. 8.5). When a single material is involved, such as a piece of steel in contact with water, then the anode and cathode are simply different areas on the surface, Fig. 8.5. The water will act as the electrolyte. The anodic area will be a region where it is slightly easier for the metal to dissociate into the ion. This can arise due to factors such as localised compositional differences, residual stresses or geometrical factors. Such factors typically generate small potential differences. More significant potential differences, and hence more problematic corrosion issues, arise when different materials are in contact with each other.

The key thing required for corrosion to occur is not that there are two different metals present, but that there is a potential difference. This voltage drives electron movement and is the driving force behind electrochemical corrosion. The potential difference can arise from two different metals being present or from two different environments being present on the same metal. As far as the electrons are concerned, it is the difference in conditions and resultant potential difference that matters, not how it arises, though dissimilar metals tend to produce a larger, and hence more significant, potential difference. When two different metals are in contact, it is the metal with the lower electrode potential that acts as the anode

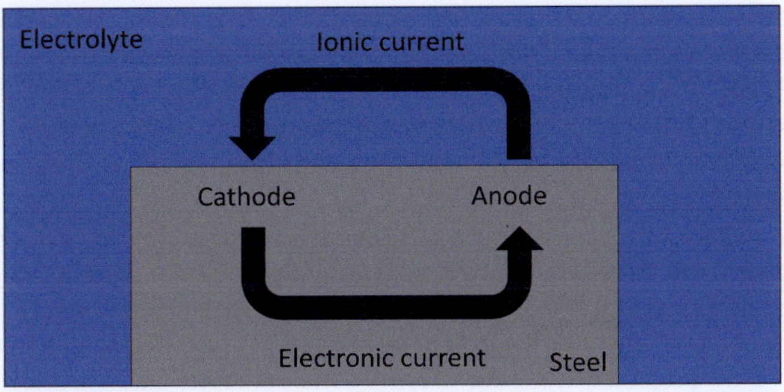

Fig. 8.5 Schematic showing key elements of the electrochemical circuit for the corrosion of steel in water. (As electrons are negative, the direction of electron flow is opposite to that of the electronic current.)

and which corrodes at an enhanced rate. Pairings of metals that have large differences in electrode potential have a large driving force for corrosion. This is why bimetallic, galvanic or dissimilar metal corrosion is such a problem. It is not just metals that can be an issue here, any conductive material needs to be considered. The potential difference between CFRP and aluminium alloys can also lead to galvanic corrosion [6].

The electrode potentials for different metals can be looked up for standard conditions, and also calculated for any non-standard conditions using the Nernst equation (Eq. 8.2). This allows the anode to be identified in any pairing and the magnitude of the driving force to be determined.

The underlying principles behind electrochemical corrosion are well known, and have been understood for over two hundred years. However, despite this, bimetallic corrosion still routinely causes extensive problems.

8.3.4 Variants of Electrochemical Corrosion

Crevice corrosion and intergranular corrosion are two variants of electrochemical corrosion where dissimilar environments result in the required potential difference. In crevice corrosion, a crevice leads to differences in chemical environments on the same component, Fig. 8.6. Water trapped in a crevice becomes depleted in oxygen preventing the oxygen consuming cathodic reaction from occurring within the crevice. In addition to generating a different chemical environment, an important aspect of crevice corrosion is that the anodic, metal dissolution, reaction is concentrated inside the crevice. Gaps a few micrometers in size are most susceptible to crevice corrosion as fluid is readily drawn in via capillary action. This results in crevice corrosion occurring where different pieces of metal are bolted together, or under rubber seals or gaskets. This also means that metal loss takes place in an inherently difficult to inspect location.

In intergranular corrosion it is the different, more open, structure of the grain boundary that provides the different environment which leads to a potential difference and corrosion. This can be exacerbated by preferential formation of precipitates in the grain boundary, which produces local differences in composition which will add to the potential difference. As shown in Figs. 8.7 and 8.8, this can lead to exfoliation in rolled material such as aluminium alloys used in aerospace when machining operations such as cutting or drilling expose grain boundaries at the cut surfaces.

Corrosion of reinforced concrete is a major infrastructure issue. Concrete loss and associated uncovering of rebars is frequently seen and is due to corrosion (Figs. 8.9 and 8.10).The first thing to understand is that it is the steel reinforcing bars that are corroding, not the concrete itself. The steel reinforcement actually starts off being protected from corrosion by the concrete. The chemical environment inside concrete is alkaline and this passivates steel. In these conditions a protective oxide layer forms which protects the underlying steel from any further corrosion. As long as the alkaline conditions persist, the steel is protected and will

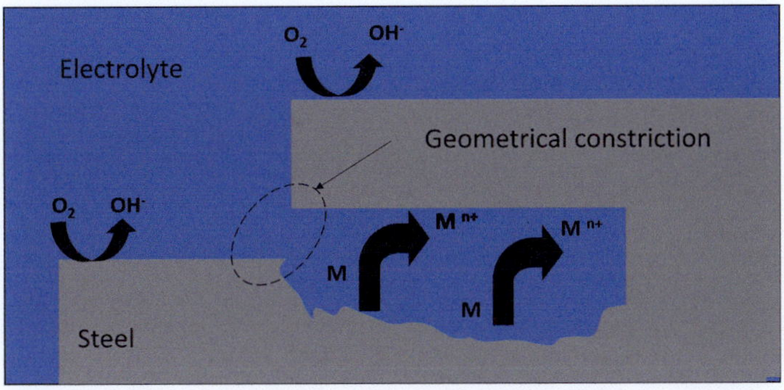

Fig. 8.6 Schematic showing how a geometrical constriction can lead to crevice corrosion by restricting resupply of oxygen to the crevice, hence limiting the cathodic reaction to areas outside the crevice with only the anodic reaction possible inside the crevice, hidden localised corrosion can result

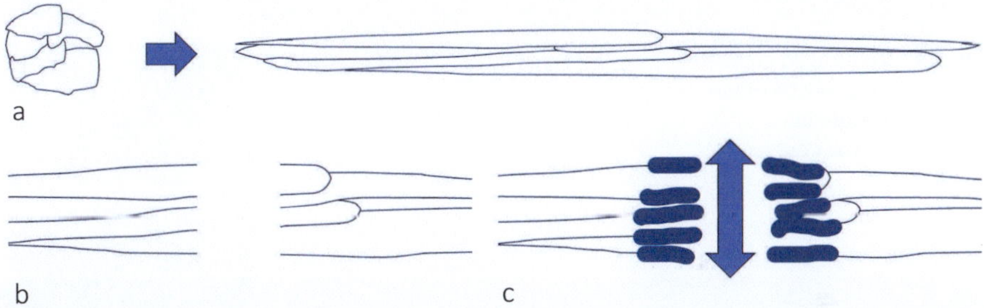

Fig. 8.7 Schematic figures showing mechanism of exfoliation corrosion. **a** Rolling processes during material manufacturing result in extended grains in thin section rolled material. **b** Subsequent hole drilling exposes grain boundaries. **c** Corrosion occurs at the grain boundaries, the volume expansion due to generation of corrosion products pushes the material apart, results in delamination of the grains

not corrode. Corrosion will only occur if the passivating oxide layer is lost. One way in which this can happen is if the internal environment becomes acidic. Unfortunately, this can happen easily. Carbon dioxide from the atmosphere diffuses into the concrete, reacts with the water that is always present within concrete, and forms carbonic acid. This changes the environment from alkaline to acidic conditions, with the conditions becoming more acidic with time as more carbon dioxide diffuses in. Once the conditions are acidic the passivating oxide is destabilised and the steel is no longer protected and corrosion starts.

A second mechanism that promotes corrosion in reinforced concrete is related to salt. In this case it is the diffusion of chloride ions that depassivates the steel allowing corrosion to occur. Relevant sources of salt are marine environments

Fig. 8.8 Intergranular corrosion produced exfoliation in this rolled aluminium alloy is clearly associated with the drilled hole. "File:Corrosión por exfoliación en aluminio–01.jpg" by Carlos Delgado is licensed under CC BY-SA 4.0

Fig. 8.9 Reinforced concrete fencing with exposed rebars due to corrosion and spallation of the concrete

and deicing salts, widely used on road infrastructure, which is largely made from reinforced concrete.

In both cases the corroding reinforcement is a problem not only due to the loss of strength of the rebars, and hence reinforced concrete, but also because corrosion products are generated and exert tensile stresses on the surrounding concrete. The concrete then cracks and can spall. This not only accelerates further corrosion but can be a significant hazard (Fig. 8.10).

Both mechanisms are accelerated when there is a thinner layer of concrete over the reinforcing bars and also when the concrete is porous. In both situations the

Fig. 8.10 Light areas on the reinforced concrete structure are where concrete spallation has occurred, in this above road location the consequences of falling concrete can be serious

amount of concrete that has to be diffused through decreases, speeding up the process. Similarly, cracked concrete will corrode more quickly as the cracks act as easy access paths for chloride ions or carbon dioxide.

8.4 Non-metallic Corrosion, in Service Degradation of Other Materials

Polymeric materials can also suffer from material degradation. Exposure to temperature can decrease the flexibility of polymers, progressively embrittling them, and induce shrinkage due to the progressive evaporation of plasticisers. This is the mechanism that produces stiffening and failure of old rubber bands (Fig. 8.11).

Polymer degradation can also be caused if conditions induce polymerisation reactions to reverse, as determined by exactly the same Gibbs free energy considerations as described in Sect. 8.3.2.1. This can be done in various ways. For polymers that are formed via condensation polymerisation the reverse reaction is hydrolysis, where water attacks the polymer, breaking bonds and cutting up the polymer molecules. For some polymers there is a ceiling temperature above which the reverse, depolymerisation reaction occurs. The polymer chain is built from many repeat units, monomers. When depolymerisation occurs these monomers detach one by one and are easily lost to the system as they are volatile. That means that they are no longer available to repolymerise if the temperature drops, reverting the direction of reaction to polymerisation, resulting in irreversible damage to the

Fig. 8.11 A rubber band which has become embrittled due to evaporation of plasticisers

polymer. In other polymers an increase in temperature can induce random chain scission, where bonds are broken at random places in the polymer chain.

The ultra-violet component of sunlight can also damage polymers as the energy supplied when ultra-violet photons are absorbed can break bonds in the polymer structure. This often results in yellowing of polymer materials and is widely seen, particularly in PVC.

The presence of heat, and or ultraviolet light, and oxygen can result in oxidation damage to polymers. This can be avoided by the use of antioxidant additives.

Polymers can also have problems related to absorption of solvents, including water. This is a particular concern in fuel systems. If there is any change in the composition of fuel, such as the UK's move to E10, petrol containing 10% ethanol from renewable feedstocks in order to decrease CO_2 emissions, compatibility needs to be checked [2]. Otherwise, polymer swelling can occur, and in extreme cases the material will dissolve.

8.5 Other Forms of Material Degradation

Stress corrosion cracking, SCC, occurs when there is the combination of a tensile stress and a corrosive environment. It is highly sensitive to the particular alloy-environment pairing. SCC is not fully understood and is an umbrella term that covers a number of different mechanisms. It is not only applied stresses, but also residual stresses that can lead to SCC occurring. Chloride environments and hydrogen sulphide, H_2S, containing environments are particularly prone to generating SCC. Copper is susceptible to SCC when in contact with ammonia containing environments. Any changes in the chemistry of the environment could lead to initiation of stress corrosion cracking in a previously corrosion free system. A review of relevant literature or dedicated tests is recommended if such changes are required.

Low temperatures embrittle materials, making them more prone to fracture. Sub-zero temperatures are a particular risk. Significant damage can be caused by ingress of water followed by freezing. The volume expansion associated with the formation of ice leads to tensile stresses in the surrounding material. This same mechanism, which can lead to fracture of forgotten wine bottles chilling in a freezer and cause fracture of water pipes in winter, can damage any material where water ingress is possible. Concrete surfaces are frequently roughened due to spallation resulting from water ingress followed by freezing. The same mechanism contributes to pothole formation in roads.

8.6 Wear and Erosion

Wear is a concern whenever moving parts are in contact. Erosion is a form of wear caused by particles in a gas or liquid impacting a surface, or by the impact of the fluid itself on the surface, particularly if the flow is turbulent. Wear and erosion can impact any type of material.

Boiler tubes are prone to fly ash erosion failure. Helicopter blades can suffer from erosion due to flying through dusty environments, the impact of dust on high velocity blades generates damage [4]. This can be a particular concern for any desert based military operations where airborne sand makes the conditions more aggressive. Similar concerns affect aircraft and wind turbines [5], both of which can be affected by water drop (rain) and particulate erosion. Surface damage to wind turbine blades has a detrimental effect on the overall energy yield [5]. Modern aircraft flight speeds are typically 900 km h^{-1}, rain impacting any material at this velocity can cause damage.

Friction, the force that resists the movement of one part over another, is closely related to wear. Friction is both material and surface topography dependent: a polished surface will have lower friction than a rough surface. It is really important to realise that the coefficient of friction is a function of the system, it is not a material property. If either material changes, or if the roughness changes, then the coefficient of friction will change. Friction is not inherently undesirable, there are many applications which need friction. Automotive tyres only propel the vehicle forward if the friction between the tyre and road surface is sufficiently high, otherwise control is lost as wheel spin and skidding occur. Trains will only move forward if there is high friction between the wheels and tracks. Leaves on the line are a big problem because they decrease the friction, the train loses traction when leaves get between the wheels and the tracks. Similarly, brake pads operate by converting kinetic energy into frictional heating, without friction these would not work. In many other cases friction is a problem. For internal combustion engine powered vehicles, it is estimated that 30% of the fuel is wasted on overcoming friction [3].

$$Q = \frac{KWL}{H} \tag{8.7}$$

The Archard equation, Eq. (8.7), is used for sliding wear. Here Q is the amount of material lost due to wear, K is a situation dependent constant, W is the normal load, L is the sliding distance and H is the hardness of the softest of the two surfaces. This gives some guidance, increasing hardness will decrease wear. Not surprisingly, decreasing load and sliding distance also decrease wear.

Lubrication works to prevent wear by preventing direct contact of the moving parts. The role of the oil film in the smooth running of pistons in automotive engines is one example of a complex system relying on lubrication. Lubricants are often liquid, as is the case with engine oil. Solid lubricants also exist, these include materials such as graphite, molybdenum disulphide and PTFE, also known as Teflon™.

8.7 Fatigue

Fatigue is the most common cause of failure in metallic materials. It is widely reported that 90% of failures in metallic materials are due to fatigue. Understanding of when fatigue is a concern is important. Fatigue can also affect ceramics and polymers.

Fatigue should be a concern whenever fluctuating stresses are encountered. Such fluctuations may have many different sources: the pressurisation cycle experienced by aircraft fuselages; thermal cycles in engines; loading cycles of large structures such as fluctuating traffic on bridges or variable wind loading. The transport industries have many components where fatigue is a concern due to the cyclic nature of the service conditions: automotive engine connecting rods, pistons, valves and cam shafts; rotating components in gas turbine engines: fan blades, compressor blades, turbine blades and the associated discs; wheels, railway lines, axles, propellors and associated transmission systems. Both regular, repeating cyclic stresses and random stress fluctuations can lead to fatigue.

Fatigue results in stress levels lower than the yield strength of the material leading to failure. The progressive growth of fatigue cracks with each stress cycle can result in failure at surprisingly low stress levels. Due to fatigue, fracture can occur at stress levels that are much lower than the constant load yield or fracture strength.

Examination of fatigue failure surfaces shows characterisation striations, each showing the microscopic crack growth associated with each cycle. Beach marks are also a feature, these highlight different growth periods (Figs. 8.12 and 8.13). These are distinctive features of fatigue failure. Under the right (perhaps that should be wrong!) circumstances, fatigue leads to incremental crack growth with each stress cycle, ultimately leading to failure when the crack exceeds a critical length.

The de Havilland Comet passenger jet crashes in the 1950's are a famous example of fatigue failure, with the fuselage pressurisation cycles being the source of the fatigue. More recent fatigue failures include the loss of the Alexander L. Kielland oil drilling rig in 1980 where fatigue cracking, related to an inappropriate weld, was the underlying cause. The fatigue was caused by fluctuating stresses from the sea. The 1998 Eschede train disaster was due to fatigue resulting from

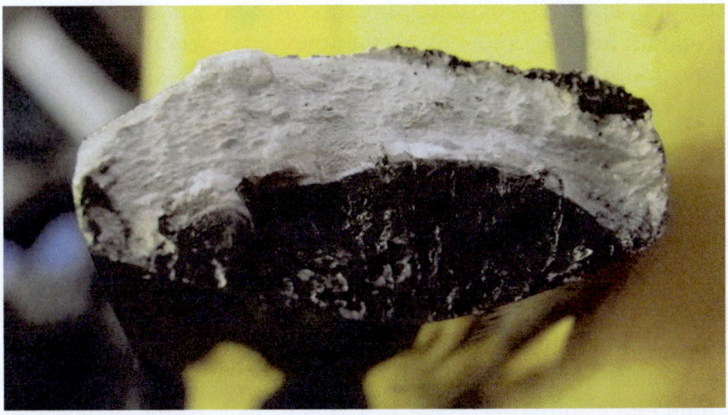

Fig. 8.12 Distinctive beach mark seen on the fatigue fracture surface of a bicycle crank. Here the cyclic stresses due to pedalling have resulted in fatigue. "Crank Fatigue Failure" by joshua_putnam is licensed under CC BY 2.0

Fig. 8.13 The fatigue failure on this bolt is seen to originate from the location of the cut thread on the right side of the bolt, beach marks indicating different periods of operation are visible on the fracture surface. "Fracture Surface" by Clint__Budd is licensed under CC BY 2.0

problems in the design of the train wheels, though there were other contributing factors. In each of these cases there were multiple fatalities.

8.7.1 Fatigue Related Calculations

Stress amplitude, S, is an important factor in fatigue, this is simply the amplitude of the relevant fluctuating stresses. This is made use of in S–N plots, where S is the stress amplitude and N is the number of cycles to failure. A typical S–N plot is shown in Fig. 8.14. It can be seen that N, the number of cycles to failure, or fatigue life, decreases as stress amplitude, S, increases. However, fatigue data is notorious for having extensive scatter, as shown schematically in Fig. 8.15. This results from the sensitivity of initiation time to surface conditions. The scatter in the data underlying S–N plots mean that they are generally plotted and used as probability contours, Fig. 8.16. The appropriate choice of the fractional probability of failure, P, will vary with application due to the different approaches to risk management.

Not all materials have S–N plots that look like Fig. 8.14. Most steels and titanium alloys have a fatigue limit, most non-ferrous alloys do not. If these materials operate at stress amplitudes below the fatigue limit then they are not affected by fatigue. Their S–N plots have the modified form seen in Fig. 8.17 where infinite fatigue life can be achieved if the stress amplitude, S, is below the fatigue limit.

S–N plots consider the overall fatigue life. However, there are actually three distinct stages in the life of a fatigue crack: initiation, propagation and final rupture. During fatigue crack initiation, stage I, a small crack forms, which then goes on to grow in stage II, propagation, before causing final rupture and failure in stage III.

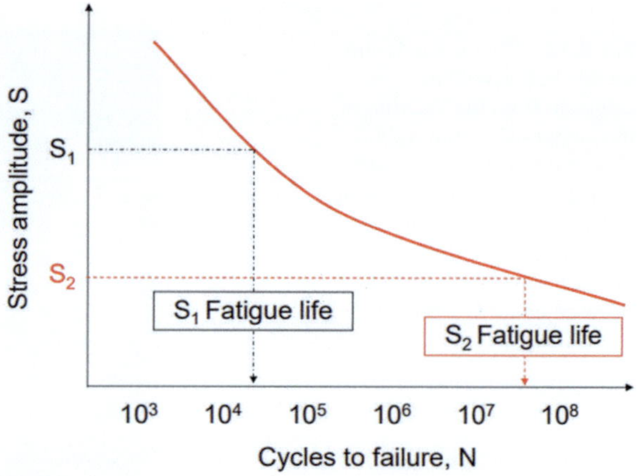

Fig. 8.14 Typical S–N plot

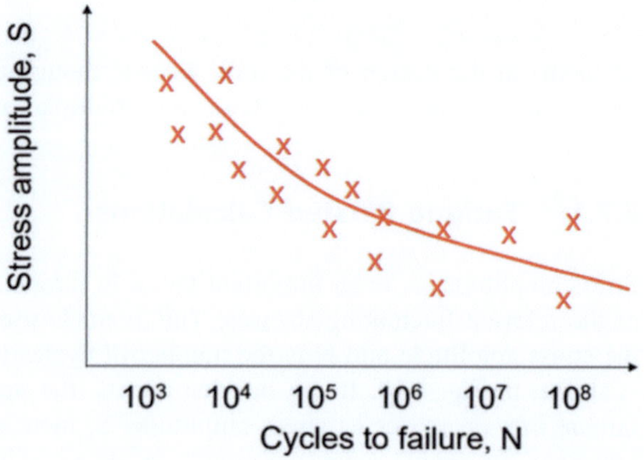

Fig. 8.15 Typical scatter in data used for a S–N plot

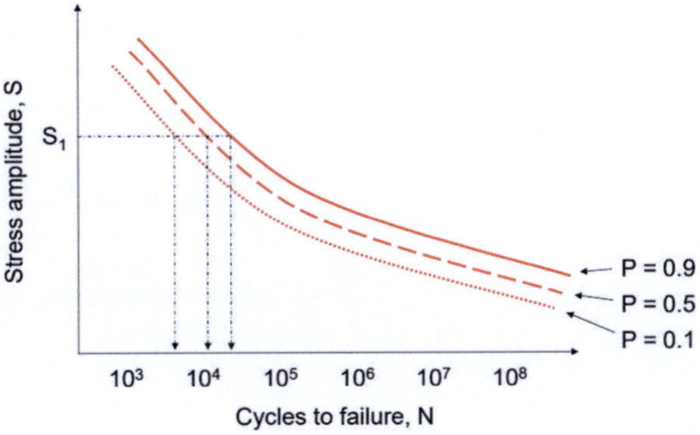

Fig. 8.16 S–N plot with probability contours. The value of P indicates the fractional probability of failure after the given number of cycles

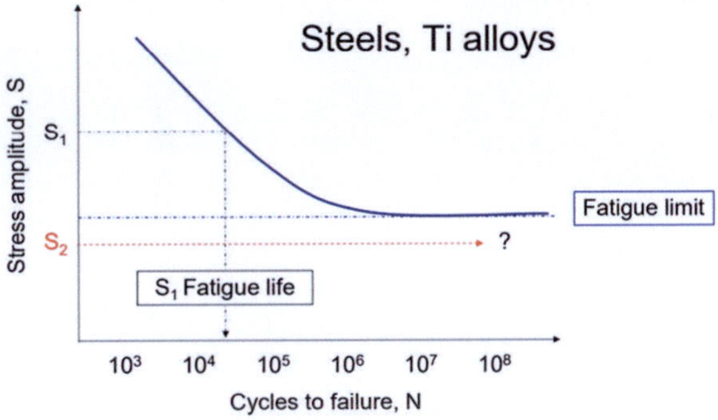

Fig. 8.17 S–N plot typical of materials with a fatigue limit such as steels and titanium alloys

In stage I, initiation, stress concentrators such as surface defects, pre-existing flaws, included phases and component geometry, locally increase the stress to levels high enough to make dislocations move. (See Chap. 12 for a full explanation of dislocations). The movement of an individual dislocation can produce a surface step equivalent to a single lattice spacing, typically about 0.3 nm. With each stress cycle more dislocations move, each adding steps to the surface. With time, the accumulated effect is roughening of the surface, the dislocation steps generate surface extrusions and intrusions. One of these will eventually get large enough to propagate as a fatigue crack.

If the surface is already rough, or contains flaws, it will take less time for a surface fatigue crack to originate in this way. Polishing the surface will extend fatigue life by decreasing the size of any surface flaws and hence extending the time for crack initiation. This sensitivity to surface details produces extensive variation in the time required for initiation and is what results in the large extent of scatter in experimental fatigue results.

Once the fatigue crack is initiated it enters stage II, propagation, and grows as shown in Fig. 8.18. The rate of crack growth increases as the crack grows, and also increases with the stress range, $\Delta\sigma$. In this stage the fatigue life can be described by Paris' Law (Eq. 8.8). As shown in Fig. 8.19, stage II comprises the majority of the fatigue lifetime. Within this stage the Paris equation can be integrated to determine how a crack will grow and how many cycles will be required to extend the crack by any given amount. A and m are constants which can be determined from the intercept and gradient of the plot respectively.

$$\frac{da}{dN} = A(\Delta K)^m \tag{8.8}$$

$$\Delta K = \Delta\sigma\sqrt{\pi a} \tag{8.9}$$

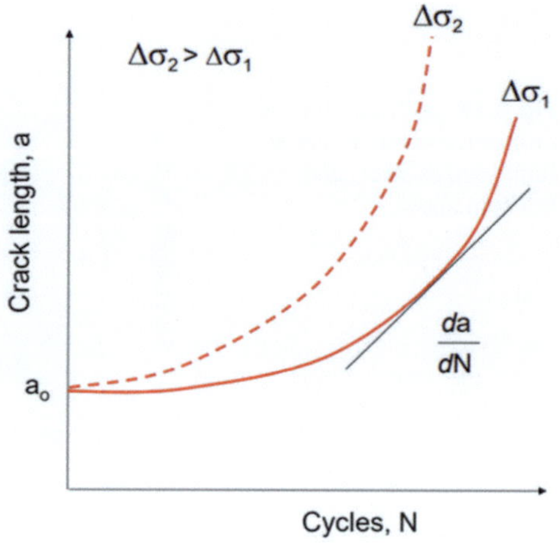

Fig. 8.18 Fatigue crack growth after crack initiation length of a_o is reached

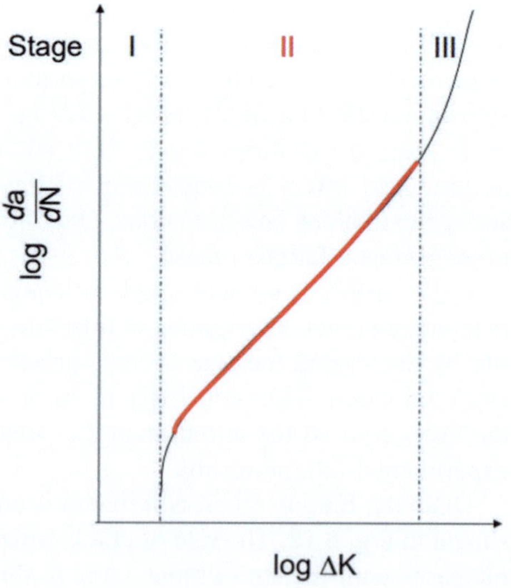

Fig. 8.19 Three stages of fatigue crack growth

Utilisation of Paris' law means that calculations can be done at inspection intervals to determine if any existing cracks in a component will grow unacceptably before the next inspection.

The fatigue risk can be minimised by eliminating stress concentrators. This extends fatigue lifetimes by increasing the time required for stage I, initiation.

Miner's law is a fractional life law. This allows the impact of changing service conditions on remaining fatigue life to be taken into account. A component has fatigue life of 10,000 cycles when run under condition 1 and a fatigue life of 5000 cycles when run under condition 2. If the component is run for 2,500 cycles under condition 1 then 25% of the fatigue life has been used up, leaving 75% of the fatigue life remaining. If conditions now change to condition 2, the more demanding regime where the fatigue life is only 5000 cycles then the remaining fatigue life is still 75%, but now that is 75% of the new conditions, i.e. 3,750 cycles. If 2000 cycles are now run under condition 2, then a further 40% of the fatigue life is used up, making a total of 65% of the life used and leaving 35% of the fatigue life of whichever future conditions are used.

8.8 Creep

Creep is of concern when materials are under load at high temperatures. Here "high temperature" means at or above 40% of the melting point when expressed in Kelvin, K. It results in progressive change of component dimensions with time. This can have massive consequences for components that operate under stress at high temperature in applications with tight tolerances. Turbine blades and automotive pistons are example applications where creep resistance is vital. Creep is most frequently associated with metallic materials but can also affect polymers. Plastic housings used to keep other components in place, such as automotive battery stacks, need to be creep resistant and maintain their dimensions in order to deliver their main function.

When a component is under load it will extend and exhibit a certain amount of strain. Usually that strain remains constant but when creep is active the strain increases with time. If the component is inspected at a later time it will be seen to have changed dimensions due to increased strain, despite the fact that the load has not changed.

As shown in Fig. 8.20, creep rates increase with both load and temperature, with the overall time to failure, or creep life, decreasing as creep rate increases.

8.8.1 Creep Mechanisms and Equations

In metals there are a number of different creep mechanisms, all of which involve diffusion. Diffusion is highly temperature dependent, resulting in creep itself being highly temperature dependent.

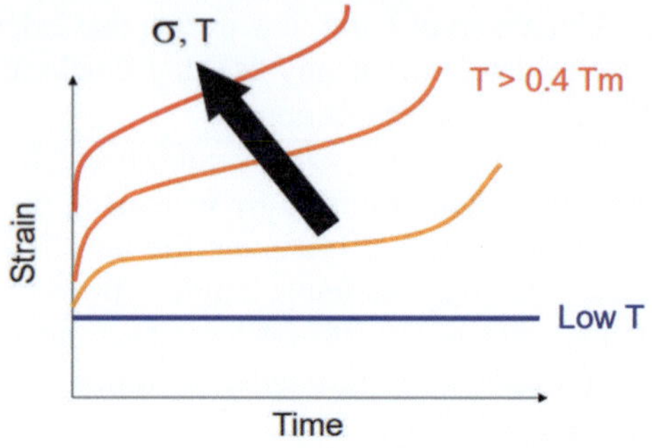

Fig. 8.20 Schematic showing variation of creep strain with time

Diffusion constantly occurs within metals, with atoms jumping randomly in different directions. The frequency of jumping increases with temperature. Under stress jumps in some directions are easier than in others. This leads to preferential movement in those directions. The resultant cumulative overall redistribution of material produces a change in dimensions. The greater the stress, the greater this effect is and hence the greater the creep rate is.

Grain boundary diffusion and bulk diffusion are two important creep mechanisms.

Grain boundary diffusion occurs in a limited volume of the overall material (Fig. 8.21). It is restricted to the disordered regions around grain boundaries where crystals, grains, of different orientations meet. These regions are more open than areas within the bulk of the material. The more open structure of the grain boundaries makes grain boundaries easy diffusion paths. The associated lower activation energy for diffusion means that grain boundary diffusion can occur at relatively low temperatures.

In bulk diffusion, diffusion occurs throughout the bulk of the material (Fig. 8.22). This material does not have the more open structure associated with the grain boundaries and hence the diffusion has a higher activation energy.

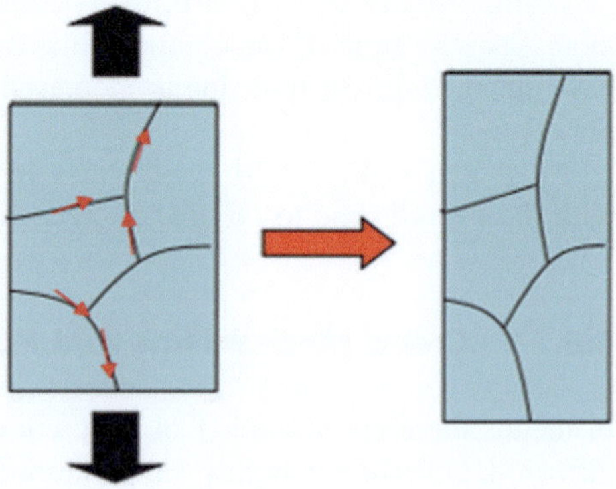

Fig. 8.21 Diffusion pathways and overall redistribution of material in grain boundary diffusion

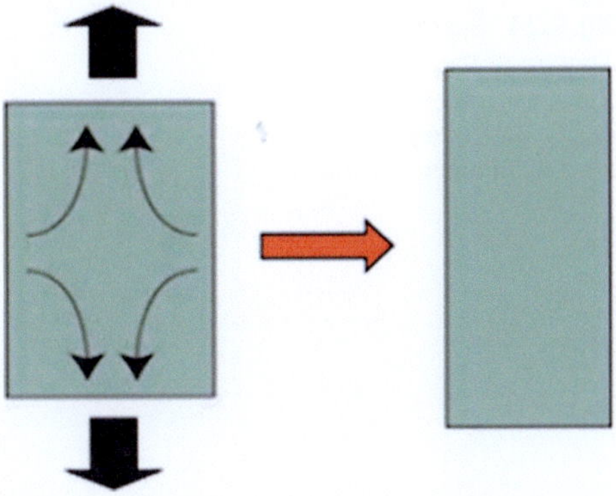

Fig. 8.22 Diffusion pathways and overall redistribution of material in bulk diffusion

Dislocation climb is another creep mechanism that also relies on diffusion. Under initial loading, dislocations will move until they encounter obstacles. Diffusion of vacancies, empty lattice sites, permits dislocations to rise up to a higher lattice plane (Fig. 8.23). This is how they can climb over obstacles.

The overall creep life of a material can be divided into three stages (Fig. 8.24). During initial loading, also referred to as stage I of creep, any mobile dislocations first move, then progressively encounter obstacles and become stuck. The final stage is stage III where the creep rate accelerates, decreasing the cross section and increasing the stress. In this stage void coalescence occurs leading to final rupture.

Stage II forms most of the creep life of most metals. The steady state creep that occurs in this stage is also referred to as secondary creep. This constant creep rate means that the creep deformation produced in a given time can be calculated. Similarly, the time that will elapse before a given strain is reached can also be

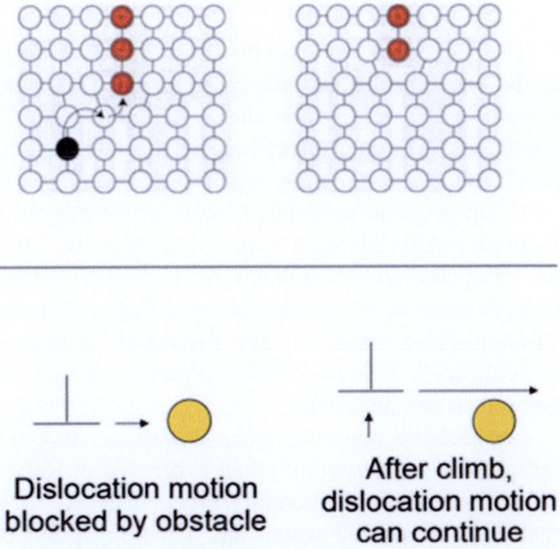

Fig. 8.23 Schematic showing how diffusion of vacancies allows dislocation climb to occur

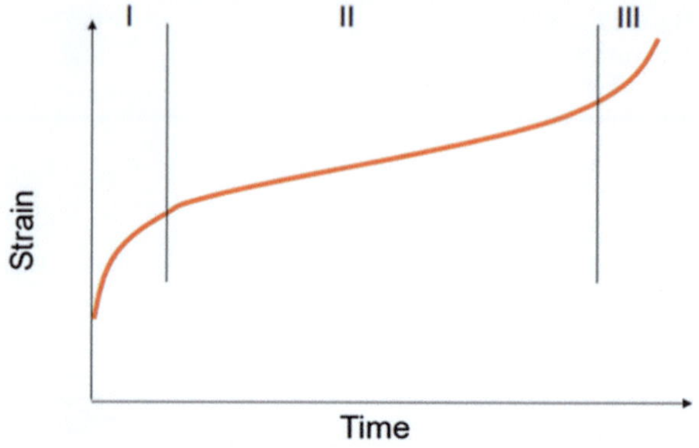

Fig. 8.24 Schematic diagram of the three stages of creep, showing that stage 2, steady state creep, dominates the overall creep lifetime

calculated. In the creep rate equation, Eq. (8.10), $\dot{\varepsilon}$ is the creep rate, $\dot{\varepsilon}_o$ and σ_o are material dependent constants, n is the stress exponent which depends on the creep mechanism, σ is the applied stress, Q is the activation energy of the relevant creep mechanism, k is Boltzmann's constant and T is the temperature in Kelvin.

$$\dot{\varepsilon} = \dot{\varepsilon}_O\left(\frac{\sigma}{\sigma_o}\right)^n exp - \left(\frac{Q}{kT}\right) \tag{8.10}$$

The creep rate equation (Eq. 8.10) can be used to quantify how changes in operating conditions affect creep. It highlights how sensitive to temperature creep rates, and hence creep lives, are. By assuming a constant strain to failure, this equation can be manipulated to produce the Larson-Miller parameter (Eq. 8.11).

$$\text{Larson-Miller parameter} = \mathrm{T}(C + lnt_r) = \frac{Q}{k} \tag{8.11}$$

This parameter varies with stress but does not vary with temperature, as long as the creep mechanism does not change. This makes the Larson-Miller parameter a really useful tool which allows extrapolation of experimental results to service conditions. The key effect here is that elevated temperature tests can be run to accelerate creep, allowing experimental results to be obtained more quickly.

Creep resistant materials have high melting points and high values of Young's modulus. Materials with large grain sizes, or, ultimately, single crystal materials are more creep resistant due to the decreased extent of grain boundaries and hence reduced amount of grain boundary diffusion. Precipitate strengthening and ordered structures also enhance creep resistance as these both make dislocation motion more difficult. Stainless steels, refractory metals and superalloys are all inherently creep resistant materials.

For metallic materials creep is only a concern when the temperature exceeds 40% of the melting point when expressed in Kelvin, K. This is why high melting point materials are inherently creep resistant. It is also why low melting point metals, such as lead which has a melting point of 327.5 °C, which is 600.7 K exhibit creep at relatively low temperatures, 40% of 600.7 K being 240.3 K, – 33 °C.

8.9 Test Your Understanding—Questions

Q2 State three different sources of potential difference that could result in electrochemical corrosion.

Q3 Using the electrode potential data in the table below determine which of each pairing of metals would be the anode, and hence actively corrode, under standard conditions. In each case comment on the relevance of the particular pairing.

a. zinc and iron.
b. copper and aluminium.

Metal	Equation	Standard electrode potential (v)
Zinc	$Zn^{2+} + 2e^{-} = Zn$	– 0.76
Iron	$Fe^{2+} + 2e^{-} = Fe$	– 0.44
Copper	$Cu^{2+} + 2e^{-} = Cu$	+ 0.34
Aluminium	$Al^{3+} + 3e^{-} = Al$	– 1.662

Q4 Name one component used in a transport application that must exhibit fatigue resistance, state the source of the fatigue related stresses and, with reference to materials properties and operating conditions, explain which material is usually selected for this application.

Q5 Describe a procedure that could be used to increase the fatigue life of a component. The description should include details of why the fatigue life is increased.

Q6

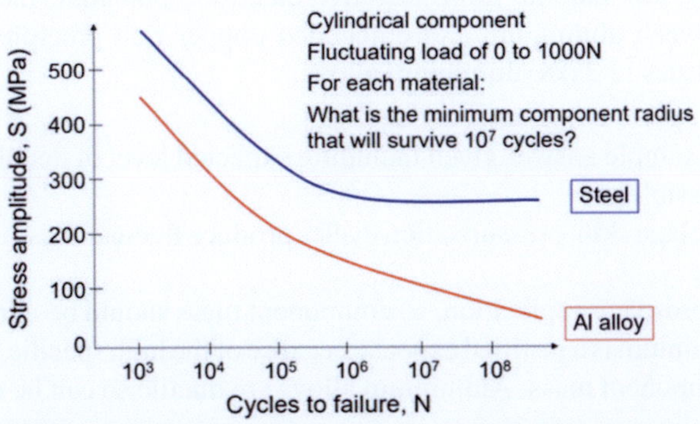

Q7 Give the equation for the Larson Miller parameter, defining all symbols used in the equation. State its use and any limitations.

Q8 A nickel based superalloy operating under a stress of 200 MPa at 750 °C has a creep life of 10,000 h. The corresponding value of the Larson Miller parameter is 32,000. Assuming that only temperature changes, use the Larson Miller equation to determine the creep life at (a) 800 °C and (b) 850 °C.

Q9 State how the fatigue life of a steel camshaft would be affected in each of the following cases:

a. The steel camshaft is laser peened before first use
b. The range of the fluctuating stresses is decreased
c. The camshaft surface is grit blasted before first use

8.10 Test Your Understanding—Answers

Q1 Answer The Nernst equation allows the effect of non-standard temperature and chemical concentrations on electrode potentials to be determined, this is required to understand what direction corrosion equations will run in under non-standard conditions.

Q2 Answer Various correct answers, three most likely are given below.
Two different metals in direct electrical contact.
Spatial variation of residual stresses within a single metal.
Spatial variation of composition within a single metal.

Q3 Answer

a. Zinc—this has the more negative electrode potential, this is also the mechanism by which galvanised steel works, with the zinc layer corroding in preference to steel if the steel is exposed
b. Aluminium—this has the more negative electrode potential, the potential difference between aluminium and embedded copper rich precipitates can cause corrosion issues in 2xxx aluminium alloys

Q4 Answer Example answer given including expected level of detail, many correct alternatives possible:

Aircraft fuselage skin, pressurisation cycles produce fluctuating stresses that make fatigue an issue.

This is an aerospace application, so component mass should be minimised, hence the cost of aluminium is a justified expense because of the high specific strength which allows low component mass. Aluminium alloys are ductile so can be readily formed into the required sheets. Not a high temperature application, aluminium alloys can operate without problems throughout the expected + 50 to – 50 °C temperature range experienced by the skin. (consideration of at least 3 different aspects of operating conditions/material propertiess required).

Q5 Answer Example answer, other correct alternatives accepted with same level of detail:

Polishing to remove/reduce size of any surface defects.

Given that a fatigue crack will only initiate with the presence of a defect that exceeds a given size any pre-existing defects give this process a head start so polishing increases the time required to initiate a fatigue crack. Increasing the initiation stage of the fatigue life will increase the total fatigue life.

Q6 Answer

Step 1—determine maximum stress amplitude by drawing a vertical line up from N = 10^7, where this intercepts the curves is the stress amplitude, this should give S = 255 MPa for steel and 100 MPa for the aluminium alloy, if you have measured slightly different values then your final answers will be a little different to what is shown here.

In each case the stress amplitude needs to be doubled to find the maximum stress—this is not always true but is true here as it is stated that the fluctuating load has a minimum value of zero.

Steel max stress = 510 MPa.

Al alloy max stress = 200 MPa.

What radius component? Stress = load/area; area = πr^2

$$\text{Stress, } \sigma = L/\pi r^2;\ r = \surd(L/\sigma\pi)$$

$$r_{st} = \surd\big(1000/(510 \times 10^6 \times \pi)\big) = 0.8\,\text{mm}$$
$$r_{al} = \surd\big(1000/(200 \times 10^6 \times \pi)\big) = 1.3\,\text{mm}$$

it should be noted that as the conditions used here coincide with the fatigue limit of the steel the radius of 0.8 mm will provide infinite fatigue life

Q7 Answer

$P = T(\ln t_r + C)$, T = temperature in K, tr = creep life, C = a constant only valid for ranges of temperatures over which there is no change in creep mechanism.

Q8 Answer

$$P = T(\ln t_r + C)$$

Need to determine t_r at a. 800 °C and b 850 °C, temperatures must be converted to K by adding 273, a. 800 °C → 1073 K and b 850 °C → 1123 K.

Rearrange LM equation: (P/T)-C = $\ln t_r$.

C is not known, but can be determined from information in the question.

750 °C → 1023 K

$$C = (P/T) - \ln t_r = (32000/1023) - \ln(10,000) = 31.28 - 9.21 = 22.01$$

This value of C can now be used to determine the creep life at different temperatures

$$\ln t_r = (P/T) - C, \; t_r = \exp((P/T) - C)$$

a. 800 °C t_r = exp((32,000/1073)– 22.01) = 2472 h
b. 850 °C t_r = exp((32,000/1123)– 22.01) = 655 h

As expected the creep life decreases with increasing temperature, the temperature sensitivity of creep is evident from the magnitude of the answers.

Q9 Answer

a. Total fatigue life extended due to extension of the initiation stage (stage 2 just shifted to later time).
b. Total fatigue life extended due to increase in length of stage 2 due to decreased stress amplitude, becomes infinite if decreased below fatigue limit.
c. Total fatigue life decreased as roughened surface decreases initiation time, stage 2 unchanged in duration.

Further Reading

Organisations and companies

The Association for Materials Protection and Performance has a website with links to many corrosion related resources https://www.ampp.org/home

The British Corrosion Institute has resources available, including galvanic series data https://www.britishcorrosioninstitute.com/

The corrosion section of the American Galvanizers Association has useful resources, including a number of examples of corrosion related failure cases https://galvanizeit.org/

Open access articles and research articles

Polymer Degradation and Stability is a journal with a number of open access articles, the journal can be searched on its website: https://www.sciencedirect.com/journal/polymer-degradation-and-stability
https://phys.org/tags/corrosion/ collates corrosion related news articles.

References

1. NACE International, March 8, 2016 "International Measures of Prevention, Application and Economics of Corrosion Technology (IMPACT)"
2. Journal of Elastomers & Plastics 2020, Vol. 52(7) 645–663 Solubility and volume swell of fuel system elastomers with ketone blends of E10 gasoline and blendstock for oxygenate blending (BOB) Mike Kass , Chris Janke, Maggie Connatser and Brian West. https://doi.org/10.1177/0095244319888773
3. How cars 'waste' two thirds of their fuel Noël Brunetière, Université de Poitiers, City Transport & Traffic Innovation Magazine March 17 2023 https://www.cittimagazine.co.uk/comment/how-cars-waste-two-thirds-of-their-fuel.html#:~:text=Moreover%2C%20not%20all%20the%20mechanical,total%20energy%20supplied%20by%20fuel, accessed 19/10/23
4. Journal of the American Helicopter Society, Vol 64, Issue 3, 2019, pp 1–5, Recent experiences of helicopter main rotor blade damage, Simone Weber, Mudassir Lone, Alastair Cooke https://dspace.lib.cranfield.ac.uk/bitstream/handle/1826/14624/Recent_experiences_of_helicopter_main_rotor_blade_damage-2019.pdf?sequence=1&isAllowed=n
5. Investigation of the impact of rain and particle erosion on rotor blade aerodynamics with an ero sion test facility to enhancing the rotor blade performance and durability J Liersch and J Michael 2014 J. Phys.: Conf. Ser. 524 012023 https://iopscience.iop.org/article/https://doi.org/10.1088/1742-6596/524/1/012023/pdf
6. Håkansson, Eva, "Galvanic Corrosion of Aluminum/Carbon Composite Systems" (2016). Electronic Theses and Dissertations. 1120. https://digitalcommons.du.edu/etd/1120

Looking After Materials in Service 9

9.1 Introduction

Surface engineering is an umbrella term for various processes which enhance the properties of material surfaces. It is the surface which interacts with the environment, hence this is where resistance is needed. Over 80% of metallic material failures are attributable to wear, corrosion and fatigue, with most of these initiating at the surface [1]. Improving the relevant properties of the surface can result in significant component lifetime extension.

The materials which provide these superior surface properties are often costly. Surface engineering ensures such materials are only used where they need to be. This combination of expensive, high performance materials with cheaper underlying engineering materials allows improved performance at reasonable cost.

Non-destructive testing, NDT, also referred to as non-destructive evaluation, NDE, and non-destructive inspection, NDI, monitors the state of components without affecting them. If deemed fit, components can then be returned to service to continue operating until the next inspection. NDT can check for deviation from component geometry, for porosity, cracks and other internal damage such as delamination. There are numerous NDT methods, with different methods being better suited to different materials and flaw types.

9.2 Surface Engineering

Surface engineering covers various different mechanisms, including mechanical and thermal processes as well as coatings and chemical surface conversion processes.

K. T. Voisey, *The Engineer's Guide to Materials*,
https://doi.org/10.1007/978-3-031-62937-2_9

9.2.1 Mechanical Processes

The related surface engineering processes of burnishing and peening are mechanical processes which are used to induce compressive residual stresses in the near surface regions of metals in order to inhibit crack opening. They also have a role in reversing the undesirable tensile surface stresses that prior mechanical machining processes may have induced in the surface. This increases fatigue lifetimes and hence extends the service life of components. Surface hardness is also increased, as is resistance to stress corrosion cracking [2].

Applications that benefit from shot peening include crankshafts, aircraft wheels, landing gear, camshafts, crankshafts and automotive gears. In short, all applications where metallic components operate in conditions of fluctuating stress, putting them at risk of fatigue failure, can benefit from peening. Rolls-Royce plc have used laser peening to treat the Ti6Al4V wide chord fan blades on their Trent engines (Fig. 9.1) [3].

Shot peening works via the impacts generated when a stream of shot, particles typically up to 0.2 mm in size, hits a surface (Fig. 9.2a). This can be done in a highly controlled way with computer controlled shot guns. Laser peening is a variant of peening where the action of intense laser pulses on the surface generates shockwaves that induce compressive stresses (Fig. 9.2b). The mechanism works by ablating a surface coating. A water layer is used to ensure the shock waves generated are directed into the material being peened. Laser peening has the capacity to induce compressive stresses to a greater depth than conventional shot peening, however it is more costly and currently a less widely established process, though

Fig. 9.1 The fan blades on the RR Trent 1000 are one of the many aerospace applications of peening. Here laser peening has been used to increase the fatigue life

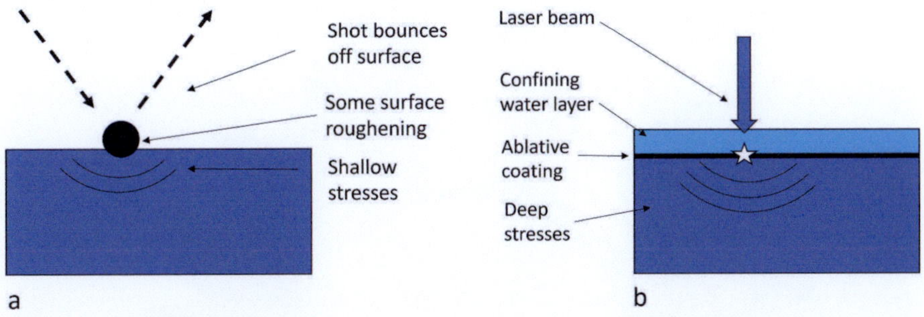

Fig. 9.2 Schematic diagrams showing **a** shot peening and **b** laser peening

the field is growing. Peening or burnishing should come at the end of the manufacturing process to avoid the desirable compressive stresses being lost due to the action of any subsequent heat treatment or other process.

These processes do not change component dimensions, though some increase in surface roughness does result, with shot peening producing more significant roughness than laser peening.

9.2.2 Surface Hardening

Thermally induced hardening of steel is another form of surface engineering which does not change material composition. Induction hardening, flame hardening and laser hardening are all based on the same mechanism. The steel is heated and then quenched, which means that the material is cooled extremely quickly. This rapid cooling is essential as it leads to a metastable form of steel forming, martensite. The carbon present within the steel gets trapped within this structure, distorting it. The resultant distorted structure is very effective at obstructing dislocation motion, making martensite a high strength, but brittle, form of steel. Slower cooling allows enough time for carbon diffusion to occur and for the crystal structure to rearrange. The high hardness of martensite makes it desirable, however the brittleness means that it often needs tempering to restore enough toughness for it to be a useful engineering material.

This mechanism is often used to increase the surface hardness, and hence wear resistance and service life, of steel gears. It is restricted to steel and there has to be enough carbon present for martensite to form, typically above 0.4%. Automotive applications include surface hardening of connecting rods, crankshafts and bearings as well as gears (Fig. 9.3). An advantage is that only the surface transforms to martensite, the underlying material remains untransformed and more ductile so will act as an effective barrier to any cracks forming in the more brittle surface. The term case hardening is also used which reflects the fact that it is only the surface that is transformed. A prior stage to diffuse additional carbon into the surface regions is frequently implemented to ensure that martensite forms. This

Fig. 9.3 Flame hardening has been applied where increased strength and wear resistance is needed, on the sprocket teeth, the rest of the material remains usefully ductile. "File: Flame hardened sprocket.jpg" by Zaereth is licensed under CC BY-SA 4.0

is termed carburising and allows the surface regions of low carbon steels to be sufficiently carbon enriched to enable martensite formation. Carbonitriding is a variant of this where nitrogen is also diffused into the surface in order to increase the hardenability of the steel, i.e. the depth to which the material's hardness is increased.

Nitriding is a related technique. Here the surface is enriched in nitrogen which then reacts to form nitrides, the presence of which hardens the surface. No quenching is needed.

9.2.3 Coatings

9.2.3.1 PVD and CVD, Physical Vapour Deposition and Chemical Vapour Deposition

The vapour deposition processes PVD, physical vapour deposition, and CVD, chemical vapour deposition, build up coatings by atomistic deposition. They are generally used to generate thin films. Coating thicknesses produced are typically 10–20 μm for CVD and 2–5 μm for PVD, meaning that surfaces can be

Fig. 9.4 TiN coated components with a distinctive golden colour. **a** TiN coated drillbit "File:Titanium nitride coating.jpg" by Peter Binter is licensed under CC BY-SA 3.0. **b** "TiN-CoatedPunches NanoShieldPVD Thailand" by Sunyataburus is licensed under CC BY-SA 3.0

enhanced with negligible impact on component geometry. PVD aluminium is used in packaging applications and PVD copper has various electromagnetic shielding applications which exploit its high electrical conductivity. There are numerous semiconductor electronic and electronic device applications, with PVD being used to deposit silicon layers as well as the transparent electrical conductor indium tin oxide, ITO, that is widely used in touch screen technology. Thicker coatings can also be produced, electron-beam PVD is used to generate ceramic top coats in thermal barrier coatings with thicknesses of approximately 200 μm.

There is widespread use of PVD and CVD hard coatings on machine tools. Materials including TiN, CrN, SiC and diamond like carbon are used alone and combined in multilayer coatings in various ways to extend the service life of many machine tools (Fig. 9.4). Titanium nitride, TiN, CVD and PVD coatings are widely used to enhance wear resistance of cutting tools, drill bits and similar applications. These give a distinctive golden appearance to the coated tools (Fig. 9.4).

9.2.3.2 Conversion Coatings

Conversion coatings deliberately enhance the naturally formed surface oxide on metal surfaces. The material is made to react electrochemically to produce the coating, as opposed to the coating being applied to the surface. This results in conversion coatings being largely free from the interfacial adhesion concerns that are usually relevant to coatings. Anodising is a well known example of conversion coating and is frequently used for aluminium alloys, generating a surface oxide layer of up to 20 μm. The thicker oxide gives improved corrosion resistance as well as surface hardness. Anodised surfaces have a porous structure. This makes them well suited to being used as an underlayer for paint as the porous surface ensures good adhesion of the outer paint layer and anodising is frequently used to enhance adhesion of paints and adhesives. The structure is also exploited aesthetically as the porous surface permits the uptake of dye, producing hardwearing coloured finishes (Fig. 9.5). Applications of anodised aluminium range from garden furniture, where the enhanced corrosion resistance and wear and fade resistant colourable finished is exploited, to handrails on the International Space Station. On the ISS the thermal insulation properties of the anodised coating are useful, these result in improved thermal stability enabling the material to better cope with differential solar irradiation. Other applications include architectural features (Fig. 9.6), food preparation equipment and automotive components.

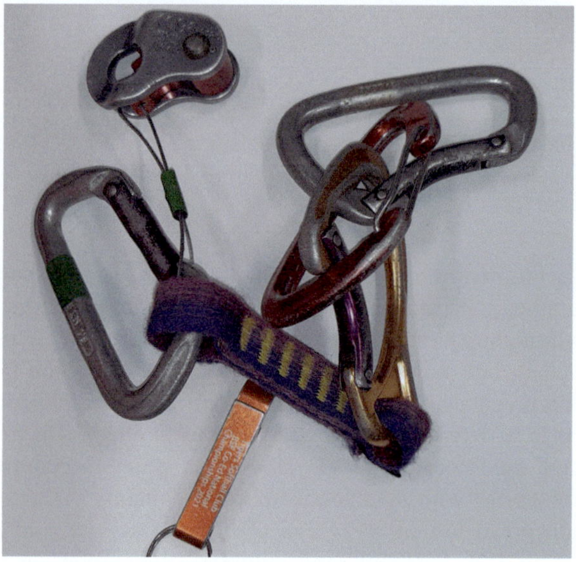

Fig. 9.5 The porosity of the anodised aluminium alloy pieces here has been exploited to colour the surfaces of these various pieces of climbing equipment and bottle opener

Fig. 9.6 Architectural use of anodised aluminium. Here the Selfridges Department Store, Birmingham, UK, makes use of the hard wearing, corrosion resistant coating to give its store a distinctive look. "Can You Count Them All?" by dolbinator1000 is licensed under CC BY 2.0

Plasma electrolytic oxidation, PEO, is a closely related technology. It its used to create hard, corrosion resistant coatings on light alloys such as titanium, magnesium and aluminium alloys. Compared to anodising, PEO uses higher voltages which produce thicker coatings, typically up to 200 μm compared to up to only 20 μm for conventional anodising. Superior corrosion and wear resistances are also characteristics of PEO coatings. PEO coatings have been applied to various applications across aerospace, automotive, biomedical, construction and the oil and gas industries. Specific applications include increasing wear resistance of titanium alloy landing gear components, being used on aerospace aluminium alloy valve bodies and actuators as well as on the aluminium alloy wheel rims of professional racing bikes, resulting in valuable weight savings [4].

9.2.3.3 Paint

Paint is one of the most widely used methods of surface engineering. It is so widely used that it almost goes unnoticed. There can be a lot more to paint than meets the eye. Frequently, the paint seen is the final, outermost, layer of a complex surface engineering system. Paint is essentially a polymer layer which works by forming a barrier between the environment and the underlying material. Paint works as long as the layer is complete, any breaches in the paint film will undermine its effectiveness as a barrier. Being a polymer means that paints have typically low maximum operating temperatures. The range of colours and finishes is achieved by additives incorporated into the paint during manufacture.

Aerospace paint applications include the decoration of commercial aircraft with corporate liveries. The continued exposure to the weather means there are durability requirements for the paint. It needs to resist impact from rain and other particulates as well as withstand continued exposure to the ultra violet component of solar radiation.

Advanced paint systems are multilayered, and can contain various additives to enhance performance. For metallic components, corrosion protection can be improved by loading the paint with particles of zinc or aluminium, significant loadings are required in order ensure the electrical connectivity required for the material to operate as a sacrificial anode. Inclusion of corrosion inhibitors also enhances performance by increasing the thickness of the naturally formed oxide layer. Corrosion protection can also be enhanced by the incorporation of inert flakes within the paint. These tend to align parallel to the surface as the paint flows during application. The flakes are impervious to water, their presence increases paint lifetime by increasing the diffusion path length (Fig. 9.7). This mechanism is successfully protecting the Forth Rail Bridge (Fig. 9.8), where utilising anti-corrosion paint which contains glass flakes has extended the paint lifetime, reducing the required frequency of painting [5].

9.2.3.4 Electrodeposition and Electroplating

Galvanised steel, where a zinc layer is added to the surface of steel to enhance corrosion resistance, is a very widely used example of surface engineering. Applications range from lampposts, railings and other outdoor furnishings (Fig. 9.9) to

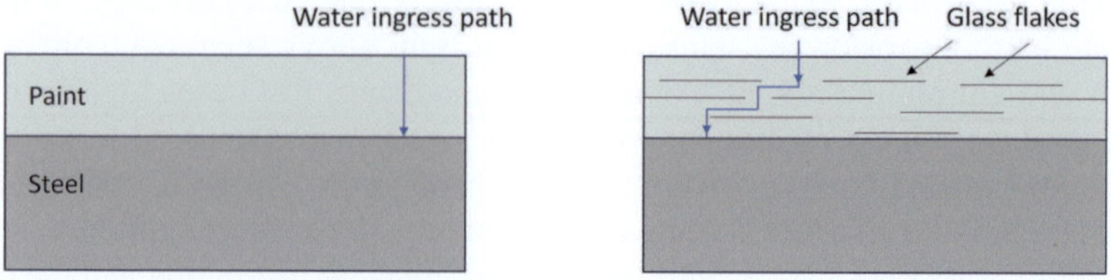

Fig. 9.7 Schematic showing how the inclusion of glass flakes increases the water ingress path length through paint, increasing the lifetime of the paint and improving corrosion protection of the underlying structure

Fig. 9.8 The Forth Rail Bridge, continuous painting is no longer required since the introduction in 2002 of paint loaded with glass flakes to enhance corrosion resistance. "Forth Bridge" by SR Photies is licensed under CC BY-SA 2.0

larger scale structures such as balconies and staircases. The automotive industry makes extensive use of galvanised steel which has had a dramatic impact on corrosion resistance, galvanised car bodies can last up to 30 years. A 10 μm zinc layer will provide approximately ten years of corrosion protection under normal circumstances. This is the technology which enables car manufacturers to guarantee corrosion resistance for up to ten years. The Renault Logan, Ford Focus and Mitsubishi Lancer are just three of a long list of car models that make extensive use of galvanised steel [6].

Galvanised steel is so effective as it acts as a sacrificial anode as well as forming a physical barrier between the underlying steel and the corrosive environment.

Fig. 9.9 Galvanised steel. The grain structure of the protective zinc layer is clearly visible on the surface of the grey square section vertical poles. The image also demonstrates how paint can be breached by in service damage

The sacrificial anode effect is due to zinc having a more negative electrochemical potential than steel. This means that if the zinc layer is breached to expose the steel, allowing an electrochemical cell to be set up between the two dissimilar metals, it will be the zinc that corrodes.

Electrodeposition and electroplating have various applications including enhancing corrosion resistance and wear resistance. There are also aerospace and automotive applications, as well decorative applications. The item to be coated is made part of an electrical circuit, an applied voltage then induces the deposition of a metallic coating. Electroless deposition produces similar coatings but uses chemical reactions to form them rather than an applied voltage. This means the substrate being coated does not need to be conductive, widening the range of materials which can be coated. This electroless deposition can be used as a preparatory stage to generate a conductive surface layer on a non-conductive material in order to allow subsequent generation of a thicker coating via electrodeposition.

Thermally sprayed coatings build up coatings which may be several hundreds of micrometers in thickness, by heating and propelling particles of the coating material towards the surface being coated. The molten, or partially molten, particles adhere on impact, progressively building up the coating. This results in a distinctive microstructure (Fig. 9.10) which can result in some differences in performance compared to bulk versions of the same material. There are a family of thermal spray coating techniques, all of which are line of sight processes. A wide range of materials can be deposited by thermal spraying onto a range of substrates. Enhancing corrosion resistance and wear resistance are common applications of thermally sprayed coatings.

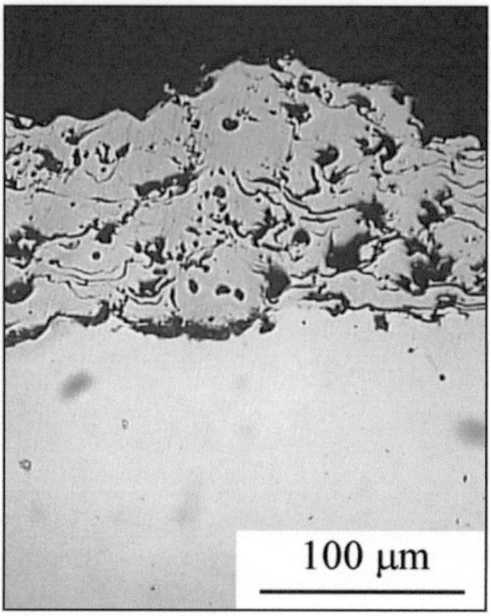

Fig. 9.10 Cross-section of a thermally sprayed NiCrAlY coating for corrosion protection. The typical complex microstructure of the coating is visible in the top half of the image

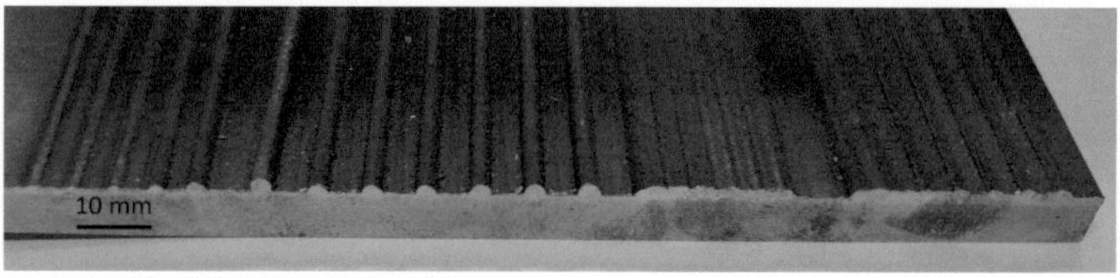

Fig. 9.11 Demonstration piece showing laser cladding of 4340 steel on mild steel to illustrate the typical dimensions and good metal to metal bonding achievable via laser cladding

Substrates surfaces do experience some heating, but this is generally less than 300 °C. An important aerospace application of thermal spraying is in the generation of thermal barrier coatings, with thermal spraying being used to deposit the outer ceramic, partially stabilised zirconia top coat. The metallic bond coat can be deposited using vacuum plasma spray, a variant of thermal spraying which avoids oxidation of metallic materials. Thermally sprayed thermal barrier coatings are used for stationary engine components such as combustion chambers, nozzle guide vanes and ducts in the hot gas path. Automotive applications include life extensions of camshafts via thermal spraying of a 13% Cr steel onto bearing surfaces.

Other applications include biocompatible hydroxyapatite coatings on orthopaedic and dental implants, where thermal spray parameters can be selected to ensure the coatings contain the porosity required to support integration with new bone growth. There are also numerous applications where thermally sprayed coatings are used in the replacement of chromium and cadmium plating, as well as in surface repair and restoration applications. Thermal spraying has also been used to enhance the wear resistance of automotive cylinder liners.

Laser cladding can be used for thicker coatings, up to several millimetres in thickness (Fig. 9.11). The laser generates a molten pool on the component surface, into which the cladding material is injected to form a surface layer. This process is frequently used to enhance the wear and corrosion resistance of metallic components. Laser cladding is part of the wider area of weld overlay coatings where welding is used deposit material on a surface, rather than as filler material. Various welding methods can be used to generate weld overlay coatings including metal inert gas (MIG), tungsten inert gas (TIG), plasma transferred arc (PTA), submerged arc welding and shielded metal arc welding (SMAW).

Automotive cylinder liners, where an inner sleeve of material is inserted into the cylinder bore to become the bearing surface, are an extreme type of surface engineering. Here a strong, wear resistant material such as cast iron is used. Inserting a cast iron cylinder liner into an aluminium alloy engine block allows the low density of the aluminium alloy to keep the overall mass low, whilst the cast iron delivers strength and wear resistance in the areas that need it. These inserts are typically a few mm in thickness.

9.3 Non-destructive Testing

The focus of this section is on actions that can be taken to look after materials in service. Non-destructive testing, NDT, permits monitoring of any progressive changes to materials due to interaction with the environment and any other components. A fundamentally important aspect of NDT is that the material is not affected by the testing. The output from NDT is data on any flaws and changes in dimensions which can be fed into fracture and fatigue calculations and lifing models. If the component is found to be sound it can be returned to service and used until the next inspection interval. NDT is also used as part of quality control processes during manufacturing. Destructive testing also has its place. It can be used in quality control where a small number of sample products are sacrificed to check that all is as it should be. There is also extensive use of destructive testing in failure analysis and forensic investigations where there can be extensive use of sectioning of materials to examine interior microstructures.

This section gives an overview of the main NDT techniques and how they are used to support the safe use of materials in service. There are many more niche NDT methods, and an indication of some of these is indicated at the end of the section. It should also be noted that industry frequently uses an array of techniques, combined into an NDT strategy, to determine if a component is fit for continued use. A good understanding of what information is required and what each method can deliver enables suitable, and cost effective, NDT strategies to be implemented.

It can be incredibly useful to record when something has been looked for and not found.

All NDT methods have their own limit of resolution, which will vary with the precise equipment used, component material and geometry. The limit of resolution is the smallest size flaw that can be detected. When using NDT it is essential to know what the relevant limit of resolution is. If the NDT results indicate that no flaws are present all it is really stating is that there are no flaws greater than the limit of resolution. There may well be smaller flaws, but as these are below the limit of resolution they simply cannot be detected. This means that you have to assume that flaws equal in size to the limit of resolution are present, and these need to be fed into any lifing calculations.

It is common practice to combine a number of different NDT techniques into an overall, multi-stage, NDT strategy. An example of this could be checking a part for which failure criteria would be surface cracks in a specified area or internal defects greater than a certain limit. Here the first stage could be visual inspection, any component with visible cracks can already be regarded as failed, with no further testing required. Only components that pass the visual inspection would then proceed to the second stage which could be ultrasonic inspection to check for internal defects. Such a multi-stage strategy can be a more cost effective approach.

9.3.1 Visual Inspection

It may be a surprise to see visual inspection listed as an NDT method. However, it would be a mistake to dismiss this technique. When used with an appropriate checklist and lighting conditions it can provide useful information, often as part of a wider NDT strategy. There is no restriction to which materials, or what component size, visual inspection can be used with. Of course, visual inspection is limited to surface features and is a line of sight process. The use of tools such as boroscopes and drones extends the line of sight and, by pushing visual inspection into regions it could not previously reach, can enable inspection without any need for disassembly. Computer aided visual inspection utilises image analysis supported by machine learning artificial intelligence systems to automate the process. The Ford Motor Company worked with IBM to deploy the IBM Maximo Visual Inspection platform, an artificial intelligence based system, to decrease production defects. The success of the system was recognised by the award of the 2021 Ford IT Innovation award.

9.3.2 Dye Penetrant

Dye penetrant testing is also referred to as liquid penetrant testing. This method is widely used to detect surface and surface connected cracks and porosity. The only limitation on what materials this can be used with is compatibility between the material and dye. It is a low cost method.

A dye is applied to the surface of the material being tested (Fig. 9.12a). Capillary action will draw dye into any surface connected cracks or porosity (Fig. 9.12b). The surface is then cleaned, removing any excess dye but leaving behind the dye that has been drawn into surface flaws (Fig. 9.12c). A developer is then applied, this is a powdery material such as chalk dust which then draws out the absorbed dye, back to the surface again via capillary action. The dye on the surface now highlights the location of any surface connected flaws and porosity (Fig. 9.12d). The technique can be enhanced by use of a ultra violet fluorescent dye which, under ultra violet light, better highlights the location of flaws (Fig. 9.13). This variant is also referred to as FPI, fluorescent penetrant inspection.

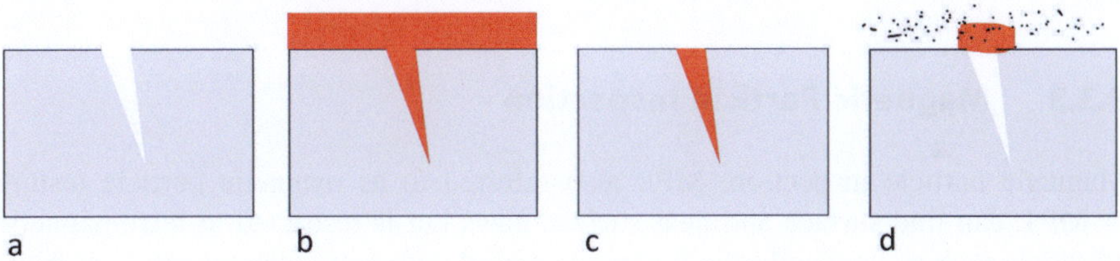

Fig. 9.12 Dye penetrant testing. **a** Item to be inspected has a surface connected crack, **b** dye is applied to surface and enters crack via capillary action, **c** surface is cleaned of excess dye, leaving dye only within the crack, **d** application of a developer draws dye out from crack, presence of dye on surface now highlights crack location

Fig. 9.13 Fluorescent penetrant inspection of a low pressure turbine casing. "FPI Inspection" by Jason Miklacic is licensed under CC BY-SA 2.0

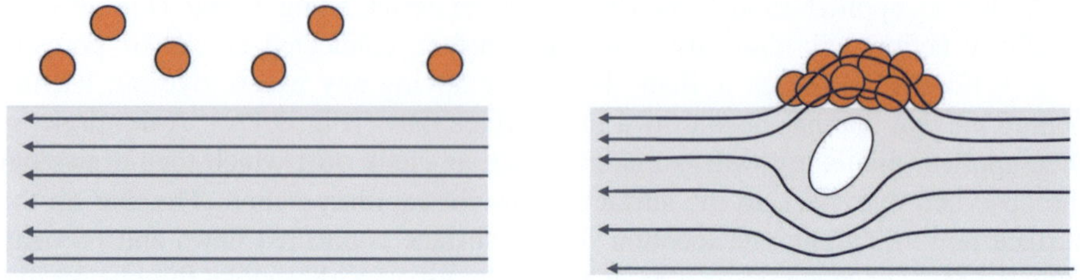

Fig. 9.14 Schematic illustration of magnetic particle testing. Left: magnetic field lines continue uninterrupted in defect free material, magnetic particles do not adhere to surface. Right: a near surface defect deviates the path of magnetic field lines, their path now passes through magnetic particles on the surface, sticking them to the surface and thereby highlighting the location of the defect

9.3.3 Magnetic Particle Inspection

Magnetic particle inspection, MPI, also referred to as magnetic particle testing or MPT, can find surface and near surface flaws but is restricted to ferromagnetic materials such as steel and iron. It cannot be used with non-ferromagnetic materials such as aluminium alloys, copper and some stainless steels. The technique induces a magnetic field in the surface of the material. Any surface, or near surface, defects deviate the magnetic field lines (Fig. 9.14). This can be detected by dusting the surface with magnetic particles, such as iron particles. The particles provide a pathway for the deviated magnetic field lines, which binds the particles to the surface. The resultant accumulation of particles near imperfections highlights the location of flaws.

Liner defects that are aligned with the magnetic field lines will not produce any significant deviation of the field lines. This means that at least two different magnetic field orientations need to be used to ensure no flaws are missed. Two different examples of magnetic particle testing are shown in Fig. 9.15, highlighting the range of component sizes it can be used for.

9.3.4 Eddy Current Testing

Eddy current testing induces electrical currents in the surface and near surface of conductive materials. The depth that can be inspected varies with material, and the precise parameters used, but is usually 4–8 mm. The technique is limited to electrically conductive materials, but can be used to measure the thickness of non-conductive layers, such as paints, on conducting materials.

The eddy current detector is an electrical coil with an alternating current flowing through it. This generates a magnetic field around the coil. When the coil is brought close to the surface of an electrically conductive material, this magnetic field induces circular, eddy, currents in the surface (Fig. 9.16). These currents then produce their own magnetic fields which in turn couple back with the detector.

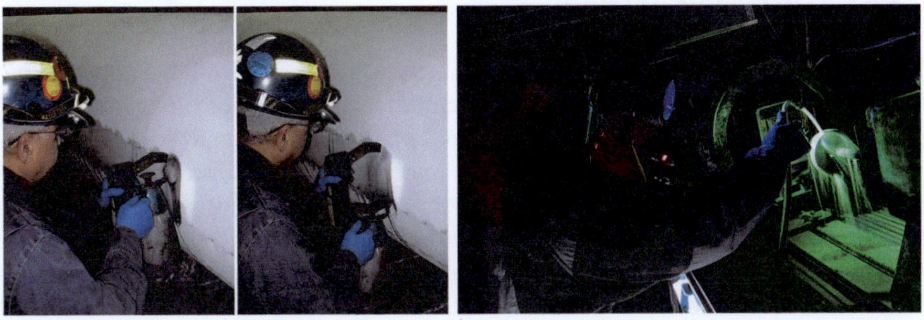

a b

Fig. 9.15 Example applications of magnetic particle testing. **a** A handheld electromagnet is used to generate the magnetic field to inspect this pipe. A black suspension of magnetic particles is sprayed onto the surface, which is painted white to enhance contrast. **b** Here the electrical coil producing the magnetic field is seen in the background, in this example UV fluorescent magnetic particles are used to enhance the process. **a** "File:Wet magnetic particle testing on a pipeline.jpg" by Davidmack at English Wikipedia is marked with CC0 1.0. **b** "170,110-F-LX370-0244" by U.S. Department of Defense Current Photos is marked with Public Domain Mark 1.0

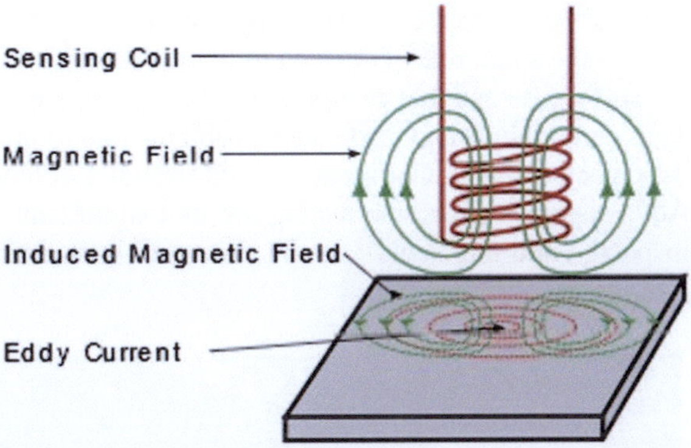

Fig. 9.16 Schematic diagram of an eddy current probe. The probe both generates and detects eddy currents in conductive materials. Defects change the eddy currents and are detected by resulting differences in the magnetic coupling with the probe. "Eddy Current" by Evan Mason is licensed under CC BY-SA 4.0

Any flaws in the surface or near surface region of the material disrupt the eddy currents which in turn change the magnetic coupling back to the detector. This can be detected by monitoring the coil's impedance variation, with any changes indicating the presence of a defect.

Eddy current testing can be used to inspect CFRP for defects and to check fibre alignment. Due to the lower electrical conductivity compared to metals, higher excitation frequencies are used in order to achieve sufficient resolution [7].

9.3.5 Ultrasonic Testing

Ultrasonic testing is a widely used technique which can be used with any kind of material. It is a pulse-echo based technique, using pulses of ultrasound to find defects. It uses exactly the same principles as medical ultrasound scans.

An ultrasonic pulse is injected into the material by a transducer, with the same transducer usually being used to detect any reflections. Any interfaces such as cracks, porosity or inclusions will generate reflections, as will the rear side of the material. The technique measures the time interval between pulse injection and receipt of the reflection. Together with knowledge of the speed of sound in the material, this is used to determine the distance of the reflection generating feature from the injection point. Multiple measurements, using different injection points are required in order to triangulate the location of reflecting defects. This also ensures that any linear defects aligned with the direction of propagation of the ultrasound are not missed.

The responses detected can be complex (Fig. 9.17), hence dedicated software and training is required. Any interfaces can generate a reflection. This includes the outer surface, where the ultrasound pulse is injected. A coupling agent is used to minimise reflection from this surface, ensuring that the ultrasound penetrates the material being inspected. The jelly put onto skin for medical ultrasound scans has exactly the same role. The need to get the ultrasound pulse effectively coupled into the surface means that components with rough surfaces can be difficult to inspect. Thin components can also cause problems due to multiple reflections complicating the detected signal. Components with irregular geometries can cause similar issues. Advantages of ultrasonic testing are that significant thicknesses of material can be inspected and that access to only one side of the material is required.

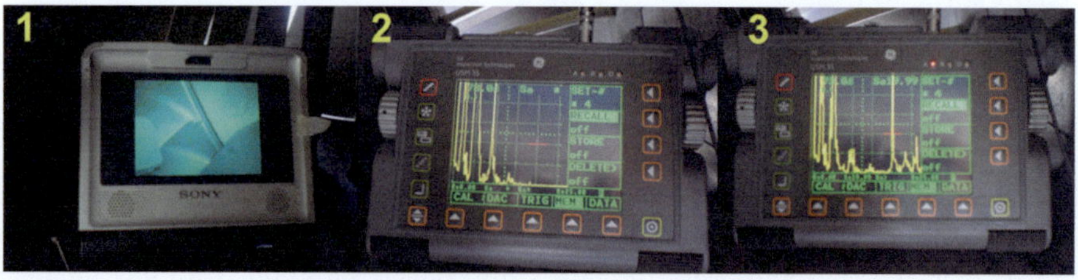

Fig. 9.17 Here ultrasonic pulse echo testing is used to examine an aerospace blade root. The area inspected is seen in Image 1. To aid interpretation, a red bar is included on screen. If a reflection to the left of the red bar, as in Image 2, this indicates a crack as the crack generates an early return echo. If the first reflection is through the red bar then the blade root is crack free. "File:NDT test of an V2500 engine blade route.jpg" by Plenumchamber is licensed under CC BY-SA 3.0

9.3.6 Radiography

Radiography uses penetrating radiation to inspect components for imperfections. It includes gamma and x-ray radiation, these are both very short wavelength, and hence highly penetrating, forms of electromagnetic radiation. Gamma rays are shorter in wavelength and are generated by radioactive decay of radiation sources such as iridium 192. The longer wavelength x-rays are produced by high voltage x-ray machines. Both x-rays and gamma radiation are hazardous, the short wavelength, high energy photons can damage biological tissue hence appropriate safety precautions must be a priority.

Both radiography techniques have the same underlying mechanism. The radiation is passed through the material to a film or detector on the rear side (Figs. 9.18 and 9.19). A proportion of the radiation is absorbed by the material, the more material the radiation passes through, the more of it is absorbed and less reaches the detector. A shadowgraph is generated at the rear side of the material, with thicker, or denser regions resulting in less radiation reaching the detector. This means that radiography essentially measures how much material is present between the source and detector. This is exactly the same principle that is used in medical radiographs. Radiography is well suited to detecting defects such as porosity, and open cracks, which change the amount of material the radiation passes through (Fig. 9.18). Closed cracks are more difficult to detect using this method due to the minimal difference in radiation path length through material.

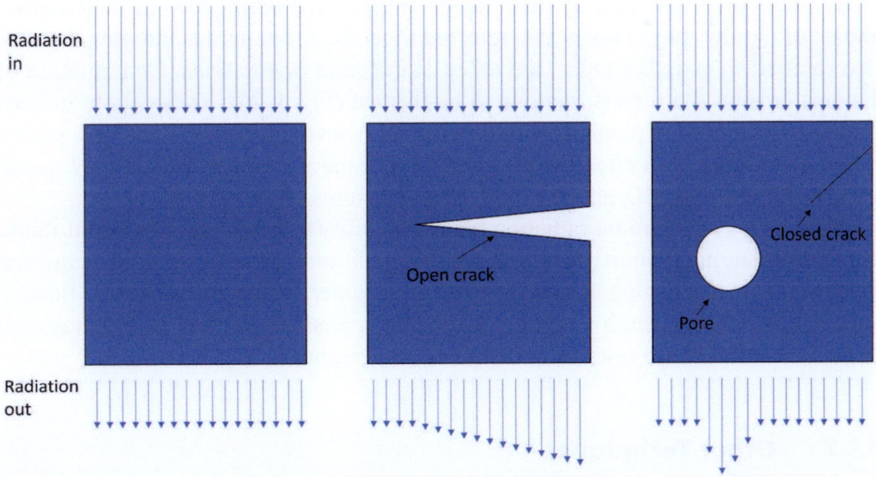

Fig. 9.18 Schematic showing how the amount of material present affect how much radiation is transmitted through an object. The more material present along the path of the radiation, the more radiation is absorbed and the amount detected on the other side is decreased. Open cracks and porosity have a measurable effect and hence can be detected by radiography, closed cracks are not easily detected as they produce minimal difference to the amount of material the radiation passes through, and hence minimal difference in the amount of radiation detected

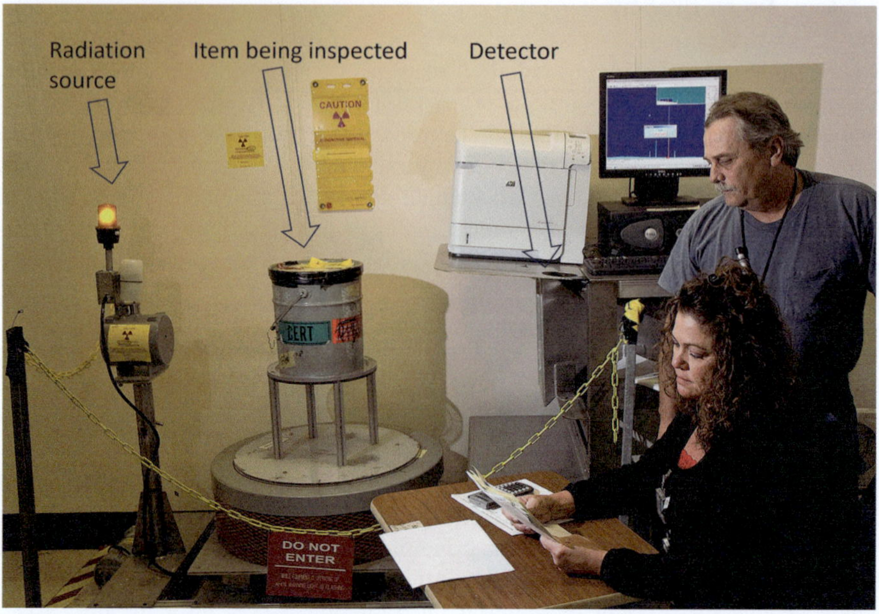

Fig. 9.19 Radiographic testing. "Non-destructive Testing at H Canyon" by Savannah River Site is licensed under CC BY 2.0. Labels and arrows have been added to the original image

CT, computed tomography, techniques can be a powerful tool to increase the impact of radiography. This is also referred to as XCT. Numerous individual scans, also referred to as radiographs, are taken at different orientations. Combination of the data enables 3D reconstructions to be formed (Fig. 9.20). These allow investigators to examine any plane within the material, moving through successive planes enable investigators to effectively move through the material which is very useful in understanding the 3D nature of relevant structures (Fig. 9.21).

Both x-rays and gamma radiation have the ability to penetrate significant thicknesses of material. Gamma radiation has the additional advantage of not requiring a power source, meaning it can be used in a wider variety of locations than x-rays, which rely on a high voltage power source. Both x-rays and gamma rays are hazardous to health so relevant precautions are required.

9.3.7 Other Techniques

The NDT techniques highlighted here are the most widely used, however there are many more NDT methods.

X-ray diffraction is used in quality control to check that single crystal turbine blades are indeed single crystals.

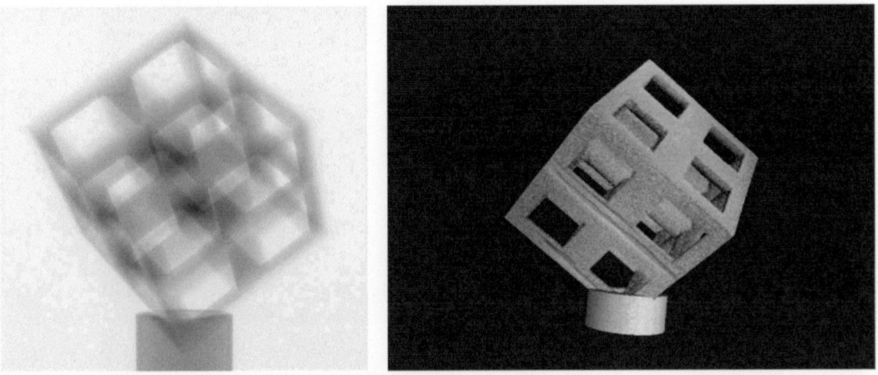

Fig. 9.20 Left: One of the 3142 individual radiographs of a nylon 12 additively manufactured part used to generate the 3D XCT reconstruction shown on the right. Edge length of the cube shaped structure is approximately 40 mm. Images courtesy of Dr Adam Thompson and the Manufacturing Metrology Team, Faculty of Engineering, The University of Nottingham

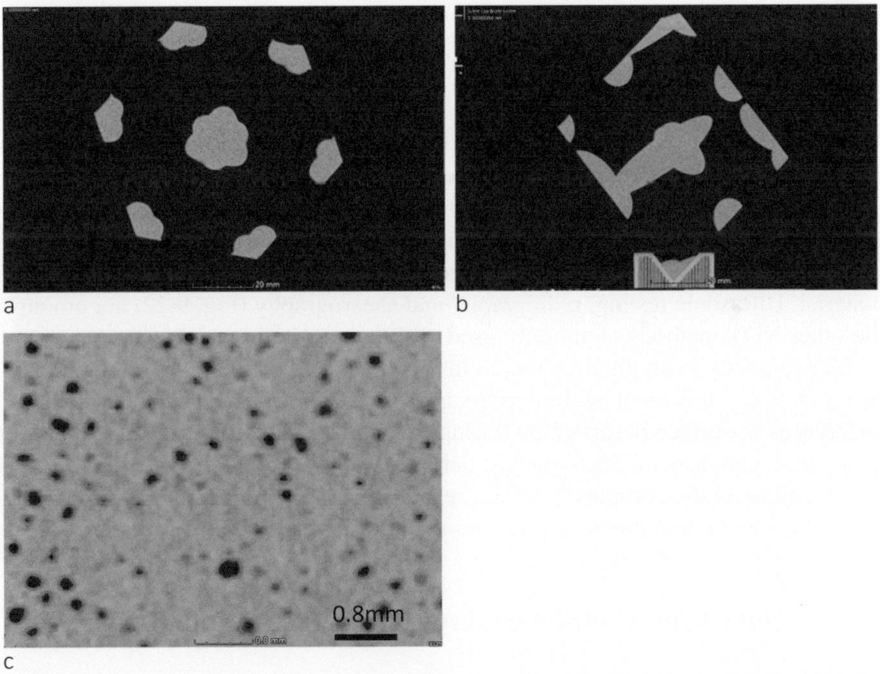

Fig. 9.21 **a, b** The usefulness of the XCT method is illustrated by the ability to extract specific planes from the reconstruction, and to interrogate specific areas, **c** a higher magnification view of embedded porosity. Images courtesy of Dr Adam Thompson and the Manufacturing Metrology Team, Faculty of Engineering, The University of Nottingham

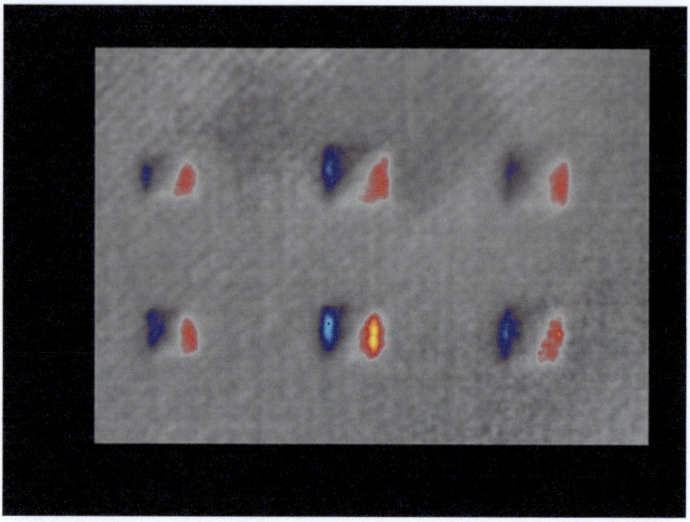

Fig. 9.22 "File:Steinbichler Shearography Honeycomb with CFRP Top Layer Artificial failures that simulate layer-core delaminations Gradient-view Result.jpg" by Shearo is licensed under CC BY-SA 3.0

Thermography uses thermal cameras to obtain images showing the temperature distribution over a surface. When used with a pulsed heat source on the rear of the material this can be used as an NDT method as the location of defects shows up as inhomogeneities in the thermograph. This is often used to inspect CFRP laminates, where any delamination will be detected as it will disrupt heat flow through the material. Ultrasonic testing, radiography and shearography (Fig. 9.22) are amongst the other NDT methods commonly used with composite materials [8].

Shearography is another technique that is made use of in the NDT of composites (Fig. 9.22). It is an optical interference based technique which detects localised differences in surface deformation during loading or excitation that discontinuities generate. It can detect defects such as debonding and delamination. A limitation of the technique is the complexity of the results generated, however it is particularly useful for large components and has no material restrictions.

9.4 Test Your Understanding—Questions

Q1 Describe and explain carburising. Give a transport application of carburising, you should state a specific component and explain why carburising is used.

Q2 A carburising process carried out for 2 h at 850 °C successfully carburises to a depth of 0.75 mm. Explain how the results of the process change if the following modifications are made:

(i) The process time was increased to 4 h
(ii) The process temperature was increased to 900 °C
(iii) The process temperature was decreased to 700 °C

Q3 Give an automotive example of the use of a coating to enhance performance of a component. Your answer should state the materials used for both the component and the coating, give a typical thickness value for the coating and explain the mechanism by which performance is improved.

Q4 Outline a non-destructive testing method that could be used to determine the flaw size distribution in an engineering ceramic. Your answer should explain the basic principles of the method and state any limitations.

Q5 A steel component can be produced with three different surface finishes:—Polished—Peened followed by polishing—Rough ground. Rank these in order of expected fatigue life and explain your reasoning.

Q6 Which of the following statements describes the capabilities of x-ray radiography?

a. good at detecting cracks, poor at detecting porosity
b. good at detecting porosity, poor at detecting closed cracks
c. can only detect crack

Q7 Which of the following NDT methods could be used to detect sub-surface porosity in aluminium alloys?

a. Dye penetrant
b. X-ray radiography
c. Ultrasound
d. Magnetic particle testing

Q8 You need to use a non-destructive testing method to check a large, approximately 50 mm deep, weld for porosity. You have access to the non-destructive testing methods listed below, but only enough time to use one of them. Select which method to use, explain why you have made this choice and comment on any disadvantages or restrictions relating to your chosen technique.

Available non-destructive testing methods:

- Visual inspection
- Radiography
- Magnetic particle testing

9.5 Test Your Understanding—Answers

Q1 Answer
Carburising is used to increase the hardness and strength of steels.

The steel is heated in a carbon rich environment to allow enrichment of the carbon concentration in the surface of the material, temperatures of 825–950 °C typically used, carbon, C, diffuses in from the C rich surface of the material—the temperature must be in the austenitic phase field. Quenching then induces the martensitic transformation where the C is trapped in the interstices of the martensitic structure, distorting the structure and thereby increasing the hardness and strength of the surface Example component, alternatives also acceptable with same level of detail: automotive gear, carburising used to increase hardness of surface in order to increase resistance to fatigue.

Q2 Answer

(i) The process time was increased to 4 h: using the approximation diffusion distance, $d = \sqrt{(Dt)}$, where D is the diffusion coefficient and t is time, $d_2 = (\sqrt{4}/\sqrt{2})d_1 = (2/\sqrt{2}) \times 0.75 = 1.41 \times 0.75 = 1.06$ mm, carburised depth is increased to 1.06 mm.

(ii) The process temperature was increased to 900 °C the increased temperature will increase the diffusion coefficient, increasing the depth to which C penetrates the surface and therefore increasing the carburised depth.

(iii) The process temperature was decreased to 700 °C temperature has been moved outside the austenitic field, hence martensitic transformation will not occur, so carburising is unsuccessful (see Sect. 5.8.1).

Q3 Answer Many possible correct answers—expected level of detail indicated below Use of galvanised steel for body panels. Body panels made from carbon steel, a layer of Zn is applied to provide corrosion resistance. The Zn layer provides corrosion protection: it acts as a physical barrier isolating the steel from the environment; the Zn is anodic to steel so if the coating is breached it will corrode in preference to the steel; 10 um of Zn is a typical thickness.

Q4 Answer Expected answer and level of detail is as below, other alternative correct answers possible.

Answer: ultrasound: An ultrasonic pulse is injected into the sample by a transducer, any discontinuities produce a reflection. The transducer detects the reflections as a function of time and hence can determine the distance of reflectors from the transducer. Measurements must be taken at multiple points to triangulate location of reflectors. Limitations: a coupling medium is needed, also the method is not well suited to rough surfaces.

Q5 Answer Shortest to longest fatigue life: rough ground, polished, peened followed by polishing. The rough ground surface will have many grinding scratches which can act as fatigue crack initiation sites, hence less time required for initiation and shorter fatigue life. Polished surface will have had surface defects removed, increasing the time required for initiation, hence increasing overall fatigue life. The surface that has been peened then polished will also have a longer initiation time, in addition to this peening will have induced compressive stresses in the near surface area which act to oppose crack opening and hence inhibit crack growth, further extending fatigue life.

Q6 Answer b. good at detecting porosity, poor at detecting closed cracks

Q7 Answer b & c, x-ray and ultrasound, (dye penetrant testing could only be used if the porosity was surface connected)

Q8 Answer Best choice is radiography—this can detect pores, visual inspection is limited to surface connected defects so would be useless to detect subsurface pores. Magnetic particle testing relies on the weld material being ferromagnetic and this information is not given in the question. Radiography is the only one of the available techniques that is definitely suitable. However, it is a high cost technique requiring appropriate health and safety measures to be put in place due to the radiation risk. Access is also needed to both sides of the weld.

Further Reading

Organisations and companies

The British Coatings Federation is the UK trade association for the sector, their website has various coatings related resources and links, including information on sustainability aspects of coatings: https://coatings.org.uk/

The International Committee for Non-Destructive Testing has NDT related resources and links to further information on their website https://www.icndt.org/.

The European Federation of Non-Destructive Testing is a member organisation aimed at supporting the European NDT community. Access to relevant standards and upcoming NDT conferences are freely available via their website: https://www.efndt.org/

BINDT, The British Institute of Non-Destructive Testing also has a number of useful NDT related links including details of various NDT training courses on their website: https://www.bindt.org/

https://www.onestopndt.com/ndt-articles/what-is-non-destructive-testing - this article is a useful overview of NDT, the main onestopndt website also has many other articles about different aspects of specific NDT methods and applications.

There are numerous companies specialising in surface engineering and NDT. Their websites typically give example applications as well as detailing the specific methods which they can supply:

https://www.eddyfi.com/

https://surecheckndt.co.uk/

https://www.flyability.com/ndt—will email a guide to NDT on submission of contact details.

https://www.cwst.co.uk/

https://www.surfacetechnology.co.uk/

https://www.oerlikon.com/balzers/uk/en/

https://www.keronite.com/about-us/

News and open access articles and research

https://www.sciencedirect.com/science/article/pii/S2212827115008021 open access review on NDT of thick composites.

References

1. Recent Developments and Novel Applications of Laser Shock Peening: A Review, Chaoyi Zhang, Yalin Dong, and Chang Ye, Adv. Eng. Mater.2021,23, 2001216 https://onlinelibrary.wiley.com/doi/full/https://doi.org/10.1002/adem.202001216
2. Shukla, P. P., Swanson, P. T. and Page, C. J. (2014) 'Laser shock peening and mechanical shot peening processes applicable for the surface treatment of technical grade ceramics: A review', Proceedings of the Institution of Mechanical Engineers, Part B: Journal of Engineering Manufacture, 228(5), pp. 639–652. doi: https://doi.org/10.1177/0954405413507250.
3. Rolls-Royce expands laser peening to include latest jet blades, 18 Jul 2013 optics.org https://optics.org/news/4/7/29
4. Xiaohu Huang, Lord Famiyeh. Plasma Electrolytic Oxidation Coatings on Aluminum Alloys: Microstructures, Properties, and Applications. Mod Concept Material Sci. 2(1): 2019. MCMS.MS.ID.000526. https://irispublishers.com/mcms/fulltext/plasma-electrolytic-oxidation-coatings-on-aluminum-alloys-microstructures-properties-and-applications.ID.000526.php
5. https://www.nms.ac.uk/explore-our-collections/stories/science-and-technology/forth-bridge-paint-mixer/#:~:text=The%20phrase%20'painting%20the%20Forth%20Bridge'%20has%20come%20to%20mean,again%20at%20the%20other%20end accessed 22/11/23
6. https://en.avtotachki.com/chto-takoe-ocinkovka-kuzova-avtomobilya-opisanie-i-spisok-modeley/#105310771082108610901086108810991077-1087108610871091108311031088108510991077-108410861076107710831080-1089-108610941080108510821086107410721085108510991084-1082109110791086107410861084 accessed 22/11/23 details of galvanized steel applications in the auto industry

7. Zhang R, Wang J, Liu S, Ma M, Fang H, Cheng J, Zhang D. Non-Destructive Testing of Carbon Fibre Reinforced Plastics (CFRP) Using a Dual Transmitter-Receiver Differential Eddy Current Test Probe. Sensors (Basel). 2022 Sep 7;22(18):6761. https://doi.org/10.3390/s22186761. PMID: 36146117; PMCID: PMC9504941. https://www.ncbi.nlm.nih.gov/pmc/articles/PMC9504941/
8. S. Gholizadeh, A review of non-destructive testing methods of composite materials, Procedia Structural Integrity, Volume 1, 2016, Pages 50–57 https://www.sciencedirect.com/science/article/pii/S2452321616000093

Materials Selection

10

10.1 Introduction

"What is the best material to use for my particular application?" is a question that is often asked, and the honest answer, always, is "it depends". The process of materials selection is carried out to ensure that there is a clear understanding of the factors on which the materials selection depends. These do include, but are not restricted to, material properties. Consideration must also be extended to the impact of geo-political events on material supply, legislation and sustainability concerns and of course cost. This means that it is just as, if not more, important to ensure that the reasoning behind materials selection decisions is recorded, as well as the final outcome.

Different companies, different designers and different engineers can come up with different answers to the same materials selection question. And these different answers can be equally valid.

Of the many other factors that are taken into account in material selection, price is the main one. This can be influenced by factors such as geographical location, the quantities of material required and the purchasing power of the relevant company. There are also considerations relating to availability of required manufacturing processes. Steel has the advantage of a well established manufacturing base whereas aluminium, though lower density, has some limitations due to the issues it has in welding due to high resistance of its surface alumina film.

This chapter is based on the systematic approach to materials selection that has been established by Prof Michael Ashby and which underlies the widely used Cambridge Engineering, CES, Selector™ software [1].

Materials indices are a powerful tool in materials selection. Materials indices are combinations of material properties that need to be maximised or minimised in order to allow candidate materials to be ranked. They can be regarded as an engineering view of material properties. Ashby plots are a widely used tool that

K. T. Voisey, *The Engineer's Guide to Materials*,
https://doi.org/10.1007/978-3-031-62937-2_10

supports materials selection via materials indices. They allow rapid comparison between candidate materials.

10.2 The Materials Selection Process

There are many applications where different materials could be used, and where there are several valid options that could be selected. An aircraft fuselage is one example: in civil aviation, aluminium alloy and CFRP fuselages are currently flying. Both materials meet the requirements of high specific strength, ability to withstand pressurisation cycles, constant exposure to weather and ability to operate in the given service temperatures. Bicycle frames are also made from various different materials, CFRP, aluminum alloys and magnesium alloys are some of the valid options, the choice between them being made based on performance requirements and price point targets. Different models of the Renault Espace have used polymer composite and aluminium alloy doors and bonnet: version 3 had polymer composite closures whereas aluminium alloy doors and bonnet were used in version 4. A demonstration of how the "right" choice for a given component can change with time as the overall design evolves.

Consideration of service conditions and material requirements are a sensible place to start regarding material selection. The automotive shell has to be able to operate in ambient weather conditions which could be wet or dry conditions, typically in the temperature range of $-$ 25 °C to $+$ 50 °C. Other material requirements are that the material is corrosion resistant, dent resistant, paintable, has minimal weight and is impact resistant.

The different materials have their own advantages and disadvantages. Materials selection is a systematic process in which candidate materials can be screened and ranked. The four stages in material selection are:

1. Translation of design requirements into a material specification
2. Screening out of materials that fail constraints
3. Ranking by ability to meet objectives, using materials indices
4. Search for supporting information for promising candidate materials

10.2.1 Translation of Design Requirements into a Material Specification

This stage of materials selection focusses on what the component is required to do. Component function, operational constraints and the design objective are specified and any free variables are identified.

Using an aircraft fuselage as an example, the operational requirement is that the fuselage has to be able to withstand an in flight pressure difference of 0.5 atm. Design requirements are the specified dimensions, for this example these are a fuselage radius of 3 m and length of 20 m. The design objective is to achieve this with the minimum mass possible.

10.2.2 Screening Out of Materials that Fail Constraints

Constraints are simply go/no go requirements that can be used to identify and remove unsuitable materials from the selection process. This includes such considerations as having a maximum operating temperature which aligns with the operating conditions, the material being compatible with the chemical environment, or being able to be made compatible via appropriate coatings etc.

For the example of an aircraft fuselage, requirements would be fatigue resistance required to withstand repeated pressurisation cycles, the need to withstand near continuous exposure to weather conditions and facilities which can manufacture the specified fuselage dimensions.

10.2.3 Ranking by Ability to Meet objectives, Using Materials Indices

The use of materials indices allows the relative merits of candidate materials to be quantitatively compared.

They are determined by extracting equations from the material requirements.

By considering the fuselage as a hollow cylinder that has to withstand a hoop stress, σ_h, where P is the pressure difference, r, the radius and t, the skin thickness, the following equation can be used:

$$\sigma_h = \frac{Pr}{t} \tag{10.1}$$

A second expression can be generated for the mass, m, of the component:

$$m = 2\pi rtL\rho \tag{10.2}$$

The skin thickness, t, features in both equations but is not specified by the design requirements, making it a free variable. The two equations can each be rearranged for t and then equated to each other to eliminate t.

$$t = \frac{Pr}{\sigma_h} = \frac{m}{2\pi rL\rho} \tag{10.3}$$

This is then further rearranged for mass, m, the characteristic that is to be minimised.

$$m = \frac{2\pi L\rho Pr^2}{\sigma_h} \tag{10.4}$$

Table 10.1 Candidate materials for consideration of use as an aircraft fuselage

Material	Density (kg m^{-3})	Strength (MPa)	$M = \frac{\rho}{\sigma_h}$	Rank
Inconel 625	8440	414	20.4	3
Ti6Al4V	4430	880	5.0	1
AA 2024-T6	2780	345	8.0	2

This shows that the component mass, *m*, is proportional to a collection of constants (2, π, L, P and r) and two material dependent terms: the density ρ and strength σ_h. The constants are the same for each candidate material and so can be ignored.

The requirement can now be rewritten as needing to minimise the materials index M where

$$M = \frac{\rho}{\sigma_h} \tag{10.5}$$

This allows different materials to be ranked according to a relevant combination of material properties.

In this example materials with the lowest value of M will be the most suitable. Different examples produce different materials indices. It should also be noted that materials indices are also referred to as merit indices.

Candidate materials for this example are shown in Table 10.1. In each case the numerical value of the materials index has been calculated, which allows the candidate materials to be ranked in order of suitability.

10.2.4 Search for Supporting information for Promising Candidate Materials

This final stage includes all other factors which may be relevant. This includes information about stability of material supply chains, availability of suitable manufacturing processes and any relevant legislation or sector guidelines. Two examples of such legislation are the End of Life Vehicle regulations which need to be taken into account by the automotive industry [2] and the European Union's Registration, Evaluation Authorisation and Restriction of Chemicals (REACH) legislation [3] which impacts all industrial sectors,

In this example Ti6Al4V has received first ranking, followed by AA 2024-T6, with Inconel 625 a distant third. Inconel 625 can therefore be disregarded. Ti6Al4V and AA 2024-T6 are kept in the running as promising materials. Further consideration of both is required to discover any relevant information or issues that also need to be taken into account in the final materials selection decision.

Ti6Al4V: high cost relative to AA 2024-T6; maximum operating temperature of 420 °C; good corrosion resistance; material supply concerns (The 2023 Russia-Ukraine conflict has impacted material supply, Russia is the third largest supplier of titanium and Boeing is reported as sourcing about a third of its titanium from Russia [4]).

AA 2024-T6: more ductile than Ti6Al4V, making manufacturing easier and less costly; potential galvanic corrosion issues with CFRP; maximum operating temperature of 160 °C.

In this example manufacturers of civil aircraft would choose AA2024-T6. Despite this coming second in the ranking to Ti6Al4V the cost difference is too great, and the difference in materials index, M, not great enough, to justify choosing Ti6Al4V over AA2024-T6.

If the example was a military aircraft where flight speeds would be great enough to cause significant frictional heating, and operational requirements may demand a higher specific strength material and cost is less of an issue then consideration of these additional factors would lead to Ti6Al4V being the material selected.

10.2.4.1 Ashby Plots

Ashby plots, also referred to as Ashby maps, are very useful tools which support materials selection via materials indices. They allow rapid comparison of large numbers of different materials to highlight which are most suitable. They are particularly effective at giving a rapid overview of which material types may be of interest.

Ashby plots have two different material properties on the axes, for the example considered here a density—strength Ashby plot is relevant. Ashby plots have log axes. This is not only to permit inclusion of a wide range of materials on the same plot but it also allows materials indices to be represented as straight lines. Materials ranking is then simplified into comparing perpendicular distances from the relevant materials index line.

The materials index expression for this example can be rearranged to

$$\rho = \sigma_h M \tag{10.6}$$

Taking logs of each side changes this to

$$log\rho = log\sigma_h + logM \tag{10.7}$$

This can be recognised as a straight line equation of the form y = mx + C, in this case y is logρ, m is 1, x is log σ_h and C, the constant, is logM. The result is that a straight line with gradient 1 has a constant value of logM at all points along the line, and hence all points along the line have the same value of M. Superposing a line with a gradient of 1 onto a density - strength Ashby plot aids material selection.

A schematic Ashby plot is shown in Fig. 10.1. Here, as is usual on such plots, different material groupings are highlighted by the corresponding areas being outlined. In addition, two specific materials are indicated by x and X. The bold line corresponds to the materials index derived for the fuselage example. All points along the line have the same value of M. The further above the line a material is

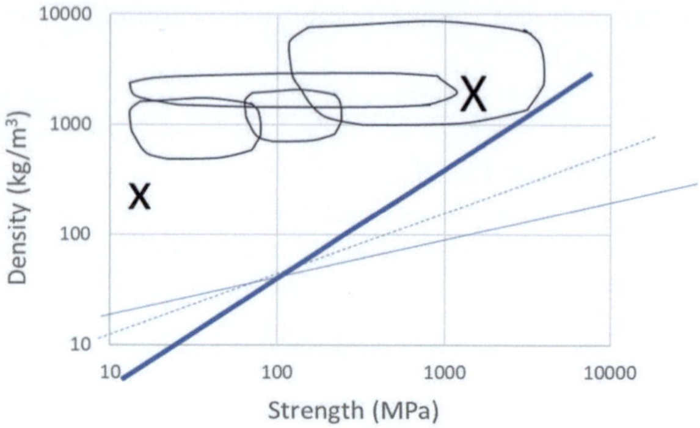

Fig. 10.1 Schematic density-strength Ashby plot. The areas corresponding to different material types are outlined. The straight lines with different gradients relate to different materials indices

the larger the numerical value of M and, in this case, the less suitable the material is. Comparing the two materials indicated by x and X it is clearly seen that x is further above the line than X, meaning that X has the smaller numerical value of M and is the more appropriate of the two candidate materials.

The thin and dotted lines have different gradients, these result from different materials indices corresponding to different requirements. It can be seen that if the materials index corresponding to the thin solid line is considered, and assuming that again the index needs to be minimsed, then in this case x, not X, becomes the preferred material.

Ashby plots, when used with material index lines, are a powerful way of rapidly selecting between candidate materials, and also of highlighting other, equally suitable, materials which may not have been initially considered.

10.3 Test Your Understanding—Questions

Q1

Material	1	2	3	4
Density (kg/m^3)	900	2800	4500	1600
Yield strength (MPa)	15	73	1100	1000
Young's modulus (GPa)	0.2	470	114	145

The four materials in the table above are candidate materials for an automotive body panel. Analysis of the design finds that the materials index $M = \rho/(E)^{1/3}$ must be minimised to minimise mass. State which of the four materials should be selected in order to minimise the mass of the automotive body panel and comment on any other relevant factors that should be considered in the final material selection.

Q2 Give one example of where legislation influences material choice.

10.4 Test Your Understanding—Answers

Q1 Answer

Material	1	2	3	4
M	1539	360	928	305

The figure of merit analysis indicates that material 4, CFRP should be selected. However current end of life vehicle legislation is relevant here as it has challenging recycling requirements which are not easy to meet if CFRP is used due to the intimate mix of two different materials, may lead to material 2, AA2024 being a better overall choice. Additional factors that must be considered are cost, material availability and corrosion resistance.

Q2 Answer Expected answers: REACH: various processes such as cadmium plating are legislated against due to environmental concerns OR ELV: end of life vehicle legislation requires vast majority of mass of car to be reused or recycled, hence manufacturers may think twice before using hard to recycle materials such as thermoset polymers.

Further Reading

Materials: Engineering, Science, Processing and Design, Butterworth-Heinemann; 4th edition (21 Dec. 2018) by Michael F. Ashby, Hugh Shercliff, David Cebon—this accessibly written book includes many examples of application of Ashby plots as well as a good overview of materials using a design led approach. One of the authors is the same Ashby that came up with Ashby plots, so this is definitely a useful resource to increase understanding of these.

From aviation to automotive—a study on material selection and its implication on cost and weight efficient structural composite and sandwich designs, M.K. Hagnell, S. Kumaraswamy, T. Nyman, M. Åkermo, Heliyon Volume 6, Issue 3, March 2020, e03716. https://doi.org/10.1016/j.heliyon.2020.e03716 is an open access research paper which uses a holistic design approach.

There are many on line resources related to materials selection, most of which feature Ashby plots and give worked examples. A few of these are listed below.

http://www-materials.eng.cam.ac.uk/mpsite/DT.html this is a very useful structured collection of resources, tutorials and other support material relating to materials selection. It includes case studies and links to material selection charts similar to Fig. 10.1.

http://www.freestudy.co.uk/nc2007%20materials/outcome%203t1.pdf an overview of different aspects of the various factors that need to be considered in materials selection, there are also useful links related to materials selection.

https://www.circulardesignguide.com/post/material-selection shows how to ensure that the circular economy is integrated into material selection.

https://www.grantadesign.com/education/students/charts/ a set of the most commonly used materials selection charts.

References

1. ANSYS GRANTA CES Selector Product Overview https://www.grantadesign.com/wp-content/uploads/2019/02/CES-Selector-2019-overview-1.pdf accessed 30/11/23
2. Statutory Instruments 2003 No. 2635, Environmental Protection, The End-of-Life Vehicles Regulations 2003 https://www.legislation.gov.uk/uksi/2003/2635/contents
3. Document 02006R1907–20231201 Consolidated text: Regulation (EC) No 1907/2006 of the European Parliament and of the Council of 18 December 2006 concerning the Registration, Evaluation, Authorisation and Restriction of Chemicals (REACH), establishing a European Chemicals Agency, https://eur-lex.europa.eu/legal-content/EN/TXT/?uri=CELEX%3A02006R1907-20231201&qid=1702299725663
4. Boeing Might Have a Russian Titanium Problem, Al Root, March 07 2022 https://www.barrons.com/articles/boeing-russia-titanium-supply-51646659011 accessed 30/11/23

Other Materials

11

11.1 Introduction

There is an almost unlimited list of other materials which have engineering applications. These include established materials, new materials and materials which have niche applications. This section highlights the particular properties of some such other materials and showcases their applications. There is also a brief overview of current areas of materials research. The intention is to serve as a prompt to be ready to think beyond the obvious when choosing materials. This may be particularly relevant in prototyping or project work where an engineer may be faced with working with a set of existing materials, the selection of which is outside of their control. This could be a scenario encountered when operating with a limited budget or when working to recycle waste material. The limitations of the various "other materials" considered here are given so that it can be understood why they are not used more widely.

11.2 Magnesium Alloys

Magnesium alloys have particularly low densities of approximately 1700 kg m^{-3}, notably lower than that of competing engineering materials such as steels, which typically have densities of around 7850 kg m^{-3} and aluminium alloys with densities in the region of 2800 kg m^{-3}. This makes them very attractive to transport industries. They also have low viscosity which makes magnesium alloys very good at filling moulds which is advantageous for complex castings. This is exploited in helicopter transmission casings, as used in the Eurocopter EC 120 and the Sikorsky S92 (Fig. 11.1). Similar applications that also make use of magnesium alloys are the intermediate compressor casings in the RR Tay and BMW/RR BR710 turbine engines, and auxiliary gear boxes in military aircraft such as the F16, Eurofighter 2000 and Tornado [1]. The automotive industry also makes use of magnesium

K. T. Voisey, *The Engineer's Guide to Materials*,
https://doi.org/10.1007/978-3-031-62937-2_11

Fig. 11.1 The Sikorsky S92 helicopter transmission casings use magnesium alloys as they are capable of producing low mass complex castings. The Sik "S92 G-WNSE IMG_6745" by Ronnierob is licensed under CC BY-SA 2.0

alloys in these kind of applications which use complex castings which are not highly loaded. BMW have used the magnesium alloy AM50 in cam covers and transmission housings where a 40% weight saving compared to steel was achieved.

In 2001 the magnesium content of a typical car was reported to be 3.5 kg, significantly lower than the 100 kg per car which was stated as a potential future outcome at the time [2]. US data from 2014 [3] states approximately 1% of vehicle mass was due to magnesium alloys. Cost, together with lack of supply chain capacity and security are the main reasons preventing further usage. However, the recent focus on energy saving and lightweighting within the automotive industry has prompted increased interest in magnesium alloys resulting in a wider range of applications, as detailed in a 2023 review paper [4].

In 2004 Aston Martin achieved a 43% weight saving by using magnesium parts in side doors. Magnesium alloy castings in door applications in the Toyota Venza and the tailgate of the 2017 Chrysler Pacifica achieved similar weight savings, as well as some part integration via redesign. Similar uses include panels in the folding roofs of Mercedes-Benz SL/SLK series vehicles and boot inner panels in the 2012 Cadillac SLS. Automotive engine blocks make use of the high thermal conductivity of magnesium, as well as the low density. Volkswagen and BMW have made use of magnesium engine blocks. BMW's N52 engine block, produced from 2004 to 2015, used magnesium alloy on the outer part of the engine, with an aluminium inner block to cope with the high loads and harsher operating conditions. Automotive gear sticks and steering wheels exploit the good mould filling ability of magnesium alloys as well as the low density. The good damping behaviour of magnesium alloys is also advantageous here.

a b

Fig. 11.2 Automotive wheels exploit the good mould filling ability of magnesium alloys as well as the good specific properties to deliver low mass wheels for high end vehicles. **a** "File:CSIRO ScienceImage 3548 TMag cast alloy wheels.jpg" by Mark Fergus, CSIRO is licensed under CC BY 3.0. **b** "Lamborghini Espada wheels" by Klaus Nahr is licensed under CC BY-SA 2.0

Magnesium alloys do not have the strength required for chassis applications but do find application in the front end carriers which hold a number of other components such as lighting and cooling systems in place. These again can be complex castings. The main advantage is again the weight saving that can be achieved, typically weight savings are about 30%. Additional benefits can be found through part integration, which will remove a number of processes, and associated costs, from manufacturing [5]. These are used in the 2012 Tesla Model S, the Porsche Panamera G2, the Range Rover and as front upper components in the 2015 Mercedes-Benz AMG GT, the 2017 Audi A8 cabin mounts. Other applications include wheels in high end cars (Fig. 11.2).

Magnesium is a highly reactive metal. This makes the raw material inherently expensive as a large amount of energy input is required to separate the metal from its ore during refinement. It also adds complications to casting as inert conditions are required to avoid any oxidation from occurring. Magnesium is well known from school chemistry lessons to be flammable. In practice, it is difficult to ignite macroscopic magnesium alloy components and it is only where there is a large surface to volume ratio, such as in machining chips or thin sheets that there is a real fire risk. The magnesium alloys Electron WE43 and Elektron 21 have passed Federal Aviation Authority flammability tests [1].

Manufacturing related complications and costs arise from the need to carry out any forming operations at high temperatures (300 °C) due to the underlying hexagonally close packed crystal structure of magnesium. There are costs due not only to the need for heating but also because steel forming equipment cannot simply be switched over to work with magnesium components. Magnesium alloys can have tensile strengths of up to about 400 MPa, which is significantly less than what can be achieved by materials such as steel. They also have a rather limited maximum operating temperature of about 200 °C, though this is alloy dependent with many magnesium alloys having maximum operating temperatures between 100 and 200 °C. In service, magnesium alloys are susceptible to corrosion and creep. This restricts magnesium alloy components to use in environments where electrochemical and galvanic corrosion is not a concern. Similarly, to avoid creep issues, magnesium alloys are not used in structural components carrying high loads.

11.3 Wood and Other Natural Materials

The drive to improve sustainability across all areas of engineering has increased interest in the use of natural materials due to the lower associated carbon emissions.

It would be a mistake to overlook wood as an engineering material. In the UK the majority of domestic roofs are supported by wood (Fig. 11.3a). The good specific strength of wood is useful here as this prevents the roof itself from adding significant additional load to the walls. The relatively low cost of wood is also advantageous as is the ease of working and established woodworking skills base.

Wood is also a current aerospace material. It is restricted to light aircraft and gliders where it has applications both as the structural framework and in the outer skin. The Pioneer 200 light aircraft has a timber framework and composite surfaces (Fig. 11.3b). The Hawk class light aircraft has plywood wing skins and the Robin aircraft has an entirely wooden structure. Wood is a natural composite, its properties are anisotropic on many levels making the overall properties of wood directional. Properties also vary from species to species with spruce, birch, ash and douglas fir being the most commonly used in light aircraft construction. Disadvantages of wood are that it can rot and the possibility of growth defects which

a b

Fig. 11.3 Engineering applications of wood **a** domestic roofing and **b** light aircraft. **a** "Roof structure" by Bryn Pinzgauer is licensed under CC BY 2.0. **b** "Fly Fano Team Alpi Pioneer 200 I-9401 1 (52237796753)" by Simon Butler from Halesowen, UK is licensed under CC BY 2.0

add variation to the material properties. Wood also absorbs water, resulting in dimensional changes with changes in relative humidity.

In the automotive industry applications of wood are restricted to interior trim, where it is used to give a veneer of luxury.

Wood has traditionally been the main material used in boat construction. The low density that enables wood to float is a key advantage here, as is the ease of working and established skills base. Wood is currently used in the construction of some small boats, with marine plywood being well suited to hull usage. Here, the use of waterproof glue prevents water ingress, allowing extended contact with water. Other marine applications include decking and superstructures on small to medium size water craft.

Engineered wood, also referred to as man-made wood, covers a number of different processes which extend the applications of wood by enhancing the overall properties. Gluing together layers decreases the impact of any imperfections, this is exploited in cross laminated timber. Densified wood has been reported to produce strengths that are on a par with steel. The various approaches used in engineered wood have combined to extend the engineering applications of wood, this has been particularly exploited in the field of architecture (Fig. 11.4).

Another engineering use of natural materials is in composites. This is an area of active research. A variety of natural fibres, such as flax, have been considered as alternative reinforcement materials in polymer composites. Bcomp®, a swiss based company focussing on sustainable composites, received significant investment from BMW, Volvo and Porsche to further work on high performance natural fibre composites [6]. Bcomp®'s material has already been used in the sunroof frame of the 2017 Mercedes-Benz E-class [6]. It is a challenge to use natural

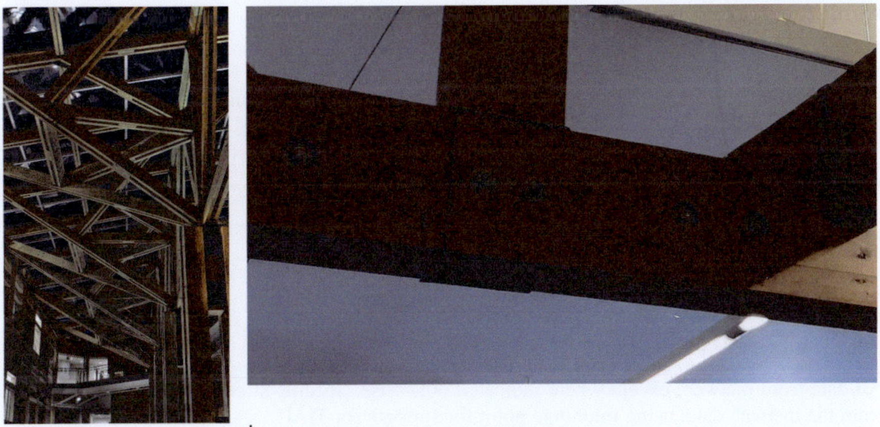

a b

Fig. 11.4 Architectural use of engineered wood **a** in the Forest Sciences Centre, University of British Columbia, **b** detail from engineered wood roof supports showing layered structure of the beams. **a** "UBC—Engineered Wood Products" by That Guy Who's Going Places is licensed under CC BY 2.0

materials in aerospace but there is active research in this area, driven by the desire to improve the sustainability of the sector. The Suspens project is investigating the use of wood as a precursor for carbon fibres [7], instead of the currently used PAN precursor which is derived from petrochemicals. Lufthansa Technik is currently developing AeroFLAX as a composite material where both reinforcement and matrix come from sustainable sources. This is expected to be used in cabin interior components rather than structural parts [7].

11.4 Intermetallics

Intermetallics are one of the materials being used to replace nickel based superalloys. Intermetallics are a group of materials that have properties in between metals and ceramics. They include Ni_3Al, NiTi and NiAl. Whilst they are made from metallic elements they are very different from metal alloys. The difference is that intermetallics have an ordered structure, not only is there a rigid ratio of constituent elements, but there is an ordered crystalline structure meaning that the constituent elements are held in a fixed 3D pattern. The ordered structure makes intermetallics inherently strong (see Sect. 4.3.1.1). This, together with their relatively low density and high melting points (Table 11.1) has made them a target material to replace nickel based superalloys. The higher specific strength of TiAl and higher operating temperatures (Table 11.1) can deliver superior engine efficiencies and reduced fuel burn. After a lot of development to overcome the issues of brittleness, manufacturing challenges and susceptibility to corrosion, intermetallics are now actively being used, lowering engine mass by replacing nickel based superalloys in some components. TiAl low pressure turbine blades are flying in the GEnx™ engine and also in PW1100G™ engines and in LEAP™ engines [8]. There is significant interest in expanding use to other applications.

Table 11.1 Comparison of key properties of the intermetallic TiAl and nickel based superalloy Nimonic 80A

	TiAl, gamma alloys [9]	Nimonic 80A [9]
Density (kg/m^3)	3910	8190
Yield strength (MPa)	400–650	780
Specific strength (MPa/kg/m^3)	0.134	0.095
Melting temperature (°C)	1460	1320–1365
Maximum service temperature (°C)	900	815 [10]

All data from matweb, except where indicated otherwise. Specific strength shown are calculated from the matweb data, using midrange point for strength for TiAl

11.5 Metal Matrix Composites

Metal matrix composites were touched on in Chap. 7 on brittle materials where the use of cermets in cutting tools was highlighted. Metal matrix composites, MMCs, have potential for much wider application. With the metal matrix usually being aluminium or titanium, MMCs have particular potential for high values of specific properties and have attracted interest across the transport industries. Boron carbide, alumina and silicon carbide are some of the materials used for reinforcement. Cosworth used a SiC/Al MMC for the pistons in their T.50 supercar [11]. Here the superior specific stiffness allowed mass reduction. The decrease in coefficient of thermal expansion, CTE, that the composite material has compared to aluminium alloys is also advantageous in this application [11]. A low CTE is also useful in space applications where differential illumination can cause problems. Space applications of MMCs include a 3.6 m long graphite/Al antenna boom for the Hubble Space Telescope [12]. Aluminium matrix MMCs were also used in the space shuttle for mid-section beams [13]. MMCs have even been to Mars, with Al/SiC being used to house the electronics for the SHERLOC tool on the Perseverance rover [14]. Automotive brake drums, gears and connecting rods have also been made from various MMCs. The high toughness of MMC fibre composites (see Sect. 6.3.2.3) has been exploited in tank armour applications. Their high specific strength has produced applications as aerospace propulsion shafts and bicycle frames.

11.6 Shape Memory Materials

Shape memory materials are more than just a novelty, they have real applications. There are shape memory metallic materials and shape memory polymers. The main application of shape memory materials is as actuators, where the shape memory material itself is usually a relatively small component but can control the positioning of attached structures. Shape memory actuators have been demonstrated in morphing wings where largescale wing shape changes result from relatively small changes in shape memory actuators. Shape memory actuators have also been employed in space where they have operated as small, lightweight deployment devices with low power requirements [15]. A particular advantage is that they can operate as low-shock release devices [16].

There are a lot of medical applications of metallic shape memory alloys such as in orthodontic braces and in coronary stents, these use nitinol, NiTi. For stents the stent is cooled and then inserted in a collapsed form, which enables easy insertion. Exposure to body temperature then triggers the transformation back to the memory, in this case more open, shape which then holds open the relevant artery.

Not all applications of shape memory materials directly utilise the shape memory effect. Shape memory alloys can exhibit the superelastic effect, where large extensions are obtained for a small increment in stress, corresponding to when the phase transformation is triggered (Fig. 11.5). Orthodontic braces exploit the

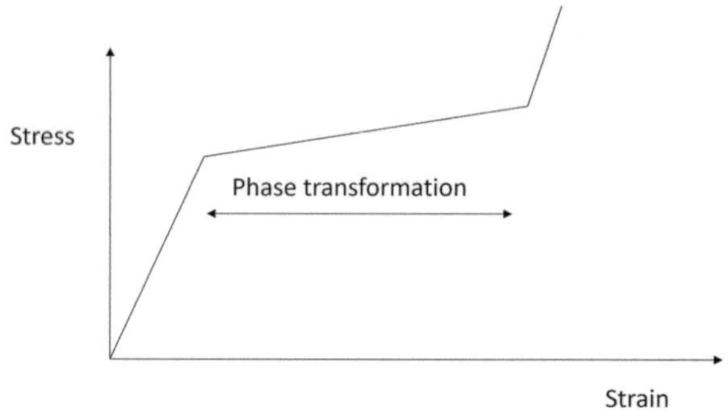

Fig. 11.5 Schematic illustration of the stress strain behaviour of a shape memory alloy. The large increase in strain for small increase in stress when the transformation occurs is the superelastic effect

superelastic effect in reverse. Body temperature triggers the transformation which is constrained by the material being fixed to teeth, hence a force is exerted on the teeth, moving them into the desired position. The advantage of shape memory alloys is that, due to the superelastic effect, a near constant force is exerted throughout the transformation, as the teeth are moved. A significant advantage of this is the reduction in the number of adjustments required to the orthodontic device while it is in use.

As is seen in Fig. 11.5, the superelastic effect produces a large area under the stress–strain graph. This area represents the amount of energy stored as the material deforms. The large size of the area shows that shape memory alloys can be very effective at absorbing energy, leading to applications that enhance impact resistance. A related feature of these materials is that they have good damping characteristics.

11.6.1 How Metallic Shape Memory Materials Work

Metallic shape memory alloys function via phase transformations. The relevant phase transformation is between a cubic crystal structure, this is the form the material is in when it is in its memory shape, and a tetragonal crystal structure. The metallic crystal structure can be regarded as being made up from millions of identical repeating cuboids (unit cells). In the cubic crystal structure these are all cubes, but in the tetragonal crystal structure the cubes distort, becoming elongated in one direction, along the x or y or z direction. The key thing is that when the material is in the cubic form there are no options, all the cubic unit cells are the same, however when it transforms into the tetragonal structure there options (Fig. 11.6).

Each cubic unit cell could distort along the x, y or z axis. Given that any macroscopic piece of material will have millions of unit cells in it, with each of these having three options as to how they distort there is an almost infinite number of configurations that the material can adopt on transformation. However, all of these will revert to the same, memory, shape when the reverse transformation is triggered.

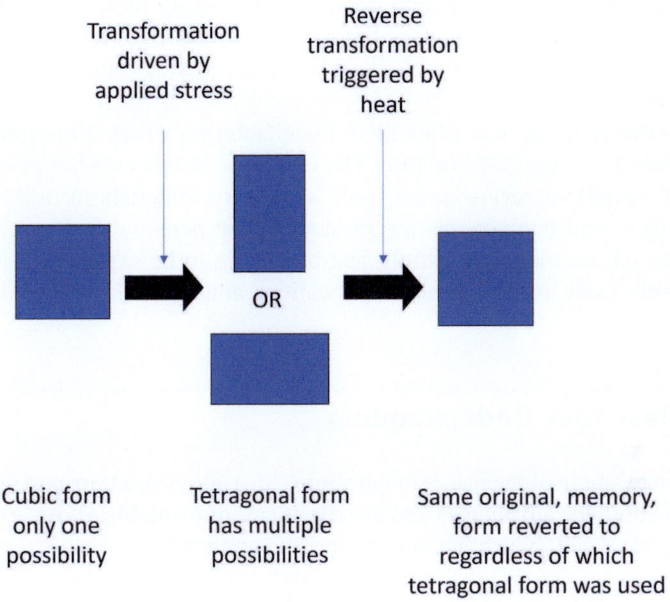

Fig. 11.6 Schematic overview of how metallic shape memory alloys work

11.7 And Also...

There are many other areas of materials research which will support future innovation and product development. According to the Web of Science™, a search engine for scientific research, there were 139,941 research publication on materials in 2022. 21,227 of these featured the word "steel", 14,777 "concrete", 10,591 "nanocomposite" and 1442 "superalloy". This is an admittedly broadbrush overview but it does demonstrate that materials research is active across the subject area. Current work includes both continuous improvement of well established materials as well as development of new materials.

Nanomaterials have huge potential. The improvement in properties that nanoscale reinforcements can, at least theoretically, provide are impressive. These however rely on effective dispersal of the reinforcement, which is inherently more challenging as the length scale decreases. Even if the reinforcement is successfully distributed it will only enhance the material if there is effective load transfer, and that requires a sound interface. This is a current challenge and there is a lot of active research in this area, including the very long list of potential applications lined up for graphene.

There is extensive research in improving the sustainability of materials, in all applications. A large part of this is related to undoing problems generated by materials themselves. Multilayered packaging is widely used in the food industry, and successfully extends shelf lives and decreases packaging mass and hence associated transport emissions, but creates big recycling challenges due to the problems of separating the materials. The challenges related to single use plastics are well documented.

There is the growing area of additive manufacturing which allows detailed tailoring of material usage and structure. The explosion in this area has pushed a need for a lot of people to review some basic aspects of materials in order to ensure mechanically sound components are produced. The potential impact of Artificial Intelligence, AI, on materials is only just beginning to be explored, it is likely to be extensively used in alloy design applications amongst other things. And also, and also…

11.8 Test Your Understanding

Q1 Give an example of the use of magnesium (Mg) alloys in a transport application. Considering both the advantages and disadvantages of using Mg alloys, explain why a Mg alloy was a suitable choice for the specific application.

Q2 Briefly describe a current area of active materials research and explain, with reference to materials properties and operating conditions, how the related advances will benefit transport.

Q3 What characteristic of metallic shape memory alloys leads to energy absorbing applications?

Q4 Give two different engineering applications of wood, in each case highlighting why wood is an appropriate choice.

Q5 Intermetallics have long been regarded as credible alternatives to nickel based superalloys. What makes intermetallics suitable as replacements for nickel based superalloys and what is the current state of progress in this area?

11.9 Test Your Understanding—Answers

Q1 Answer Many possible correct answers—example answer given:

Application, e.g. automotive steering wheel or dashboard.

Advantages: good mould filling allows castings with surface details to be produced; low density Disadvantages: cost related to specialised processing needed. In the specific example of an automotive steering wheel the good mould filling ability is exploited in the generation of this complex shape. The low density is also beneficial to decrease the total vehicle mass. The advantages outweigh the additional cost and complexity of manufacture, making an Mg alloy a suitable choice. Nb both advantages and disadvantages must be listed to get full marks.

Q2 Answer Example answer given below, many possible acceptable alternatives, marking scheme below indicates expected level of detail: Development of refractory alloys to replace superalloys. The high temperatures and high loads that turbine blades experience are very demanding operating conditions. Further increase in operating temperature is desired in order to improve engine efficiency however any significant future increase in operating temperature will require moving to a different material type, one with a higher melting point such as refractory alloys. Alloy development is required to produce refractory alloys that have the required combination of mechanical properties, temperature capability and corrosion resistance that would balance their higher densities and successfully improve engine efficiencies.

Q3 Answer The super elastic effect.

Q4 Answer Many possible answers, the two most likely are given below.

Light aircraft—wood frequently used for the main structure, can also be used for wing surfaces, the low density, good strength parallel to the grain mean that low mass structures are produced, also the material is readily available and there is a widely established manufacturing base.

Domestic roofing support structures—the low density, good strength parallel to the grain mean that low mass structures are produced, which minimise the additional load that the roof generates on the structure, the ease of joining and shaping to form complex frame structures is also an advantage.

Q5 Answer The high maximum operating temperatures and low densities are attractive, the yield strengths are less than those of nickel based superalloys, however lower mass components can be produced because intermetallics have higher specific strengths. Sufficient progress has been made in intermetallics that they are now in service as low pressure turbine blades in gas turbine engines.

Further Reading

Organisations and companies

The Institute of Materials, Minerals and Mining (IOM3) is the relevant UK professional body, its website https://www.iom3.org/ has numerous materials related resources as well as news articles on the latest developments in the area.

The European Materials Research Society has links to materials related events as well as freely accessible articles on different topics within materials on its website https://www.european-mrs.com/

Open access articles and research papers

The extent of current research in materials is challenging to keep up with. There are many dedicated journals: Wikipedia lists about 120 materials science journals. The move to greater open access publication means an ever growing number of materials research publications are freely available to all. Frontiers in Materials is an open access journal with articles available on all types of materials, including review articles which can be a useful starting point to understanding the current state of the art of any specific materials topics https://www.frontiersin.org/journals/materials. Other materials journals tend to have a proportion of articles available via open access, including "open access" in any web search on the topic of interest is advised in order to ensure you can access articles found.

References

1. Bruce Gwynne and Paul Lyon, Magnesium Alloys in Aerospace Applications, Past Concerns, Current Solutions, Presented at the Triennial International Aircraft Fire & Cabin Safety Research Conference October 29 - November 1, 2007 https://www.fire.tc.faa.gov/2007Conference/files/Materials_Fire_Safety/WedAM/GwynneMagnesium/GwynneMagnesiumPres.pdf
2. Magnesium Casting Applications in the Automotive Industry, Roderick J. Esdaile, SAE Transactions Vol. 110, Section 5: Journal of Materials and Manufacturing (2001), pp. 381–385 https://www.jstor.org/stable/44699793
3. FY 2014 Annual Progress Report – Lightweight Materials R&D, U.S. Department of Energy, Energy Efficiency & Renewable Energy, 2014 Vehicle Technologies Office https://www.energy.gov/sites/prod/files/2015/07/f24/DOE%20VTO%202014%20Materials%20Annual%20report.pdf
4. Development and application of magnesium alloy parts for automotive OEMs: A review Bo Liua, Jian Yang , Xiaoyu Zhang , Qin Yang , Jinsheng Zhang , Xiaoqing Li, Journal of Magnesium and Alloys Volume 11, Issue 1, January 2023, Pages 15–47
5. Magnesium Car Parts: A Far Reach for Manufacturers? Part 1,Gary Kardys | November 25, 2017, insights.globalspec.com
6. https://www.fiberjournal.com/natural-fibers-the-new-fashion-in-automotive-composites/#:~:text=The%20principal%20natural%20fibers%20being,considered%20for%20specific%20end%2Duses. International Fiber Journal, Natural Fibers: The New Fashion In Automotive Composites, Geoff Fisher February 20, 2023
7. https://aviationweek.com/aerospace/emerging-technologies/bio-sourced-materials-could-make-composites-greener Aviation Week Network Bio-Sourced Materials Could Make Composites Greener, Thierry Dubois February 13, 2023
8. B. P. Bewlay, S. Nag, A. Suzuki & M. J. Weimer (2016) TiAl alloys in commercial aircraft engines, Materials at High Temperatures, 33:4–5, 549–559, https://doi.org/10.1080/09603409.2016.1183068
9. matweb.com
10. https://www.specialmetals.com/documents/technical-bulletins/nimonic-alloy-80a.pdf
11. MMC Pistons for the T.50 Supercar, Gary S. Vasilash, Published 2/16/2021, composites worldhttps://www.compositesworld.com/news/mmc-pistons-for-the-t50-supercar

12. M. Badiey, A. Abedian, Application of Metal Matrix Composites (MMCs) in a Satellite Boom to Reduce Weight and Vibrations as a Multidisciplinary Optimization , ICAS 2010 27TH International Congress of the Aeronautical Sciences https://www.icas.org/ICAS_ARCHIVE/ICAS2010/PAPERS/144.PDF
13. https://pure.manchester.ac.uk/ws/portalfiles/portal/261216005/FULL_TEXT.PDFNext-Generation Metal Matrix Composites for Space Launch Applications, Luke J. Rollings, PhD Thesis, School of Natural Sciences, Department of Materials, The University of Manchester 2022
14. https://cpstechnologysolutions.com/one-year-ago-cps-lands-on-mars/ accessed 22/11/23
15. https://technology.nasa.gov/patent/LEW-TOPS-135 Mechanical And Fluid Systems, Shape Memory Alloy Mechanisms for CubeSats (LEW-TOPS-135), Lightweight and efficient mechanism for retention, release, and deployment of solar arrays and antennas
16. https://pubmed.ncbi.nlm.nih.gov/32326510/ Costanza G, Tata ME. Shape Memory Alloys for Aerospace, Recent Developments, and New Applications: A Short Review. Materials (Basel). 2020 Apr 15;13(8):1856. https://doi.org/10.3390/ma13081856. PMID: 32326510; PMCID: PMC7216214.

Dislocations

12

12.1 Introduction

Dislocations are fundamental to making metals useful. Metals are useful because they are strong and tough enough to form useful, robust, components. Without dislocations metals would be much stronger, but they would also be incredibly brittle and would behave like glasses, with the smallest defect easily growing into a crack causing fracture. Dislocations give metals their malleability and ductility, the properties that are key to being able to shape them into useful components.

Saying that metals are useful is of course a massive understatement. Our ability to manufacture items by changing the shape of metals underlies all technological progress. Knowingly, or unknowingly, we have been manipulating and exploiting dislocations for about 5000 years.

Dislocations are microscopic defects in crystal structures, they can also occur in some ceramics but it is in metals where they are most important. The word defect implies that they are undesirable, this is not the case. Dislocations are always present in metallic materials used for engineering applications. There are two exceptions to this. These are metallic glasses and all metallic materials when lowered to the temperature of absolute zero. Metallic glasses are only achieved when metals are cooled from the melt very, very, very quickly. It is challenging to generate bulk quantities of metallic glasses. Their properties are different to normal, crystalline, metallic materials. Currently there are only a small number of specialised applications, but this is growing and the field is an active area of research. The temperature of absolute zero, − 273.15 °C, is impossible to reach. The closest achieved to date is 38 trillionths of a degree above it [1], and that was done under highly controlled conditions which are not going to be encountered outside of world leading physics laboratories. Together this means that any engineers who deal with metallic structures and components are working with materials containing, and relying on, dislocations. This makes it worth investing a little time to understand what they are.

K. T. Voisey, *The Engineer's Guide to Materials*,
https://doi.org/10.1007/978-3-031-62937-2_12

A lot of the materials science of metallic materials is really centred on controlling the motion of dislocations.

12.2 What Are Dislocations?

Dislocations are a geometrical defect in the crystal structure. Figure 12.1a shows a perfect region of the crystal structure, and Fig. 12.1b shows an equivalent region which contains a dislocation. The dislocation is highlighted by black atoms. It can be seen that the top four rows of atoms each have an extra, black, atom present. Noting that what is shown in Fig. 12.1 is a cross section through the material it can be understood that a dislocation can be regarded as an extra half plane of atoms inserted into the material. The crystal structure in the immediate vicinity of the dislocation is distorted in order to accommodate the dislocation.

Between Fig. 12.1b and c the dislocation has moved to the edge of the sample, producing a surface step. The intermediate steps are shown in Fig. 12.2, highlighting the progressive nature of the motion. As the dislocation moves through the material, as shown in Fig. 12.2, the atoms in the top half of different columns in turn become part of the dislocation. It is important to note that while the dislocation itself travels from the middle to the edge of the material, individual atoms move much smaller distances. A frequently used analogy is of a ripple moving in a rug. While the ripple passes from one side to the other, each part of the rug will have only moved by a small amount. Where the ripple in the rug results in the overall displacement of the rug, the passage of the dislocation results in the displacement of the top of the material with respect to the bottom.

The ripplelike motion of the dislocation allows the overall motion to occur progressively. This is far easier, requiring far less force, than if the same motion occurred in one step. This is what gives metals their malleability, which allows them to be relatively easily formed and shaped. Without dislocations metals would be both far stronger, requiring higher forces for deformation, and also far more brittle, being more likely to fracture rather than yield, overall behaving much more like glass. With dislocations, layers of atoms can slide past each other at usefully low stresses. This dislocation motion is called yield. Yield is required for permanent, plastic, deformation of the metal to occur.

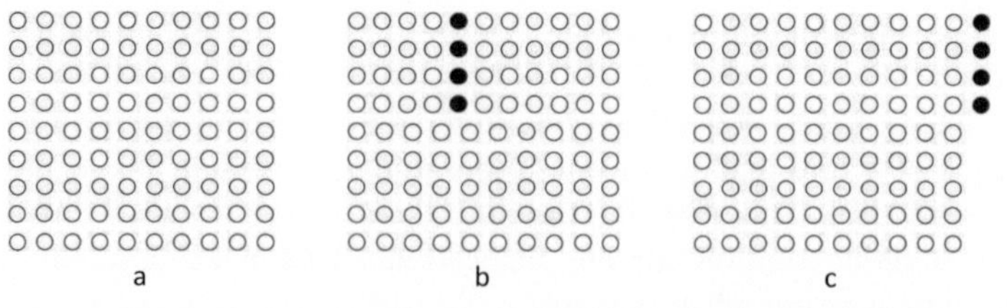

Fig. 12.1 **a** Dislocation free crystal structure, **b** dislocation highlighted by black atoms and **c** surface step formed by motion of dislocation

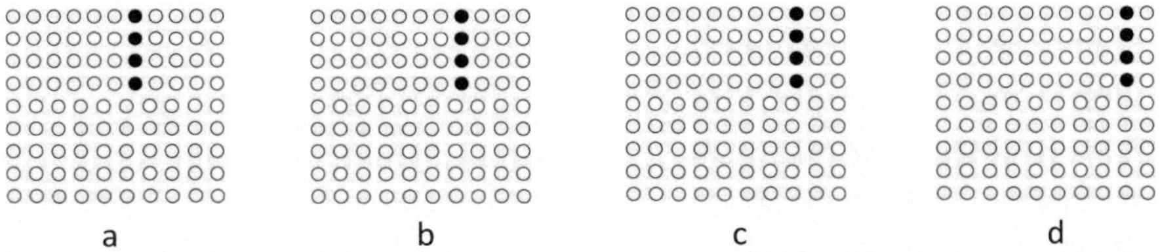

Fig. 12.2 **a–d** shows the intermediate positions of the dislocation as it moves from the position in Fig. 12.1b to that in Fig. 12.1c

Making dislocation motion more difficult strengthens the material by making yield, and hence plastic deformation, more difficult. Dislocations move most easily in perfect crystals with no imperfections. Dislocation motion can be impeded by anything that disrupts the regular crystal structure. Obstacles to dislocation motion include alloying elements, grain boundaries, precipitates and other dislocations.

12.3 Controlling Dislocation Motion

Controlling dislocation motion is the underlying aim of a lot of work done by materials scientists and metallurgists. This is done by manipulating the obstacles to dislocation motion within the material. Understanding the strain field that dislocations induce in materials is key to understanding how and why they interact with different microstructural features.

Dislocations are often represented by an upside down T symbol, where the stem of the symbol represents the extra half plane of atoms. The cross of the T represents the lattice plane that the base of the dislocation is on. By considering the need for the crystal lattice to have to distort to accommodate the extra half plane of atoms associated with the dislocation, it can be recognised that there must be regions of compressive and tensile strain induced in the crystal structure in the region immediately around the dislocation. Figure 12.3 is a schematic highlighting this. The region above the lattice plane is in compression as it has to distort to accommodate the extra material. This compressive strain has to be balanced by a tensile region, the region of tensile strain is the region below the dislocation. This tensile region can be thought of as having to stretch to accommodate the mismatch in number of lattice planes above and below the dislocation.

Due to these regions of strain there is a certain amount of stored elastic strain energy associated with dislocations. This means that increasing the number of dislocations requires energy input, and also that there is a thermodynamic driving force to decrease the number of dislocations, as that will decrease the overall energy of the system. The amount of this stored energy scales with the strain squared. This is really important in understanding why dislocations act as obstacles to other dislocations.

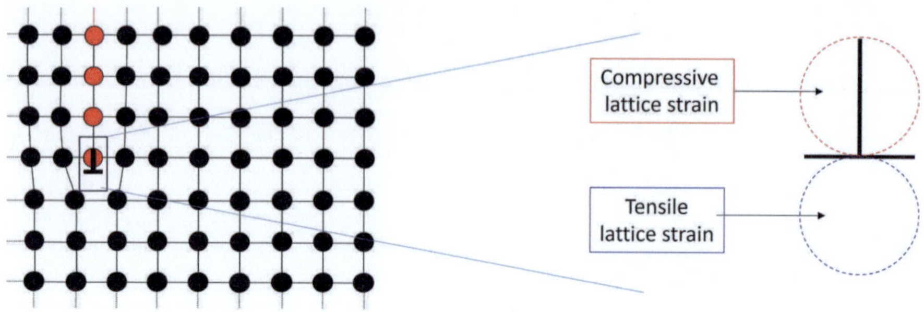

Fig. 12.3 Schematic of the regions of induced strain around a dislocation

As two dislocations of the same type approach each other their strain fields start to overlap. The compressive region of one overlaps with the compressive region of the other, and similary the tensile strain regions also overlap. The strain at any point is then the sum of the strain from the first dislocation plus the strain from the second dislocation at the same point (Eq. 12.1). However, the stored energy at any point is more that the addition of the strain energy associated with each dislocation individually (Eqs. 12.2 and 12.3). This is because the elastic stored strain energy is proportional to the total strain squared and the square of the total strain is always greater than the total of the individual squared strains (Eq. 12.2). The closer the dislocations get, the more they overlap, the more points have increased total strains, and the greater is the increase in stored strain energy. This means it requires effort, energy input, to supply the additional stored strain energy and move the dislocations closer to each other. This is how dislocations can act as obstacles to each other.

$$\varepsilon_{TOT} = \varepsilon_1 + \varepsilon_2 \tag{12.1}$$

$$(\varepsilon_1 + \varepsilon_2)^2 > \varepsilon_1^2 + \varepsilon_2^2 \tag{12.2}$$

$$S_{TOT} = S_{(\varepsilon_1+\varepsilon_2)} > S_{\varepsilon_1} + S_{\varepsilon_2} \tag{12.3}$$

The effect of solid solution strengthening is similar a solid solution is when alloying additions exist as individual atoms within the host crystal structure, as opposed to being incorporated into embedded particles. The difference in size between the alloying element and the main element means that the crystal lattice is distorted, with larger alloying elements causing greater distortion, Fig. 12.4. Here it can be seen that the large blue atom has distorted the surrounding lattice, putting it into compression to accommodate the larger element. If the dislocation now traverses the material, due to an applied stress, it will move from left to right horizontally across the material. In this particular example, the tensile strain region below the dislocation will overlap with the compressive strain region around the

Fig. 12.4 Schematic showing lattice distortion associated with substitutional solid solution alloying elements

blue atom. There will be some overall reduction in strain as the compressive and tensile strains cancel each other out, and a corresponding decrease in the total stored strain energy. The more the tensile strain region of the dislocation overlaps with the compressive region of the alloying element the greater this decrease in energy will be. At the point of maximum overlap, and minimum stored strain energy, the dislocation can be regarded as sitting in an energy valley. Moving it further will require additional effort due to the associated increase in stored strain energy as overlap decreases, and total strain increases again. This means that the dislocation sticks at the minimum energy point. Additional effort, i.e. increased applied stress, is required for further motion, which is the same as saying that the yield strength has increased.

Precipitates are particles of a different phase embedded in the main crystal structure. These can cause a significant disruption to the crystal lattice, and hence are another feature that can act as an obstacle to dislocation motion. When thinking about the interactions of precipitates and dislocations it is usual to view dislocations in a different way. So far cross-sections of dislocations have been considered. The viewpoint will now be changed to look down on a dislocation from above (Fig. 12.5). From this perspective the dislocation is seen as a line, this can be regarded as the top of the extra half plane in the crystal structure.

As seen in Fig. 12.5, as the dislocation meets precipitates its progress through the material is interrupted. As the dislocation continues to move forward under the applied stress only part of it can move whilst part is pinned by the precipitates. This causes the dislocation to bow, increasing the total length of the dislocation line. Due to the strain fields around the dislocation there is a stored elastic strain

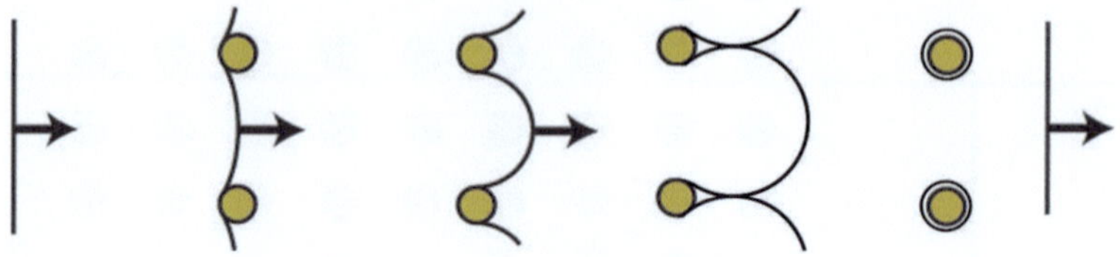

Fig. 12.5 From left to right, a dislocation line moves through a material, encounters precipitates which obstruct its motion, pinning the dislocation and requiring it to bow through the precipitates. With sufficient bowing, the dislocation line joins up with itself, resulting in orphaned dislocation loops left behind around the precipitates as the dislocation line continues to move through the material

energy associated with each unit length of the dislocation. As the dislocation line gets longer, the total stored strain energy increases. The associated required energy input is what makes moving a dislocation through a distribution of precipitates more difficult than moving through precipitate-free material. Eventually the dislocation bows round so much that it bows round on itself, with different parts of the dislocation line coming into contact with each other. The dislocation line then gets pinched off, leaving orphaned dislocation loops around the precipitates as the line continues to move forward through the material (Fig. 12.5).

Grain boundaries, the interfaces between adjacent individual crystals which form the metallic structure are significant disruptions to the crystal lattice and as such act as obstacles to dislocation motion (Fig. 12.6). Smaller grain size metallic materials are higher in strength than larger grained versions of the same material. This is due to the increased number of grain boundaries per unit area, and hence larger number of obstacles to dislocation motion, present in fine grained materials. This effect, the Hall Petch effect, is quantified via the Hall Petch equation (Eq. 12.4), where σ_y is the yield strength, σ_o, is a material dependent constant

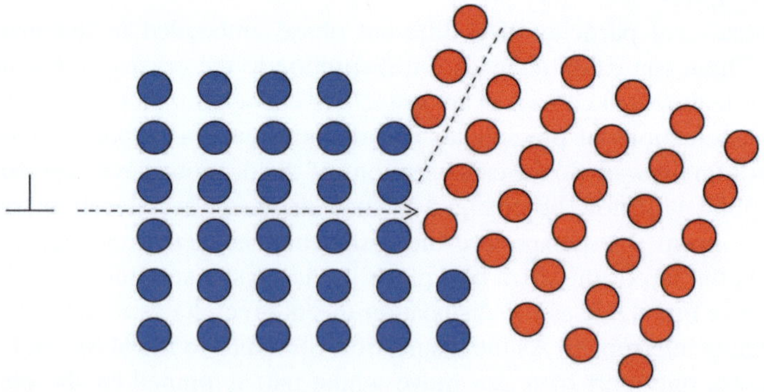

Fig. 12.6 Schematic showing the significant disruption to dislocation motion caused by grain boundaries, the boundaries between the individual crystals which form the metallic material

equivalent to the lattice strength, k, is a material dependent constant and, d, is the grain size. This equation works for grain sizes down to about 10 nm.

$$\sigma_y = \sigma_o + kd^{-1/2} \tag{12.4}$$

12.4 Annihilation, Recovery and Recrystallisation

Dislocations are always present in metallic materials. There are processes which can change the number of dislocations, the dislocation density. Dislocation sources are features in metallic materials which generate additional dislocations when dislocations interact with them. This means that any operation involving plastic deformation of a metal, such as rolling, forging, hammering etc., results in an increase in the dislocation density. Dislocations act as obstacles to other dislocations. The increased dislocation density following deformation results in a higher yield strength due to the increased number of interactions. This is the phenomenon of work hardening.

The increased dislocation density means that the total elastic stored strain energy has also increased, and that, in line with thermodynamics, changes that will decrease this will be favoured. A material with a high dislocation density will rearrange itself to decrease the dislocation density if possible. Thermal activation energy is what makes this possible, by allowing diffusion and hence dislocation motion to occur.

Annihilation is a thermally activated process which decreases the total number of dislocations. In annihilation, two opposite dislocations approach each other. As they become close their strain fields overlap, but because these are two opposite dislocations, the compressive strain field of one will overlap with the tensile strain field of the other (Fig. 12.7). There is a resultant decrease in total stored strain energy as the tensile and compressive strains cancel each other out. The greater

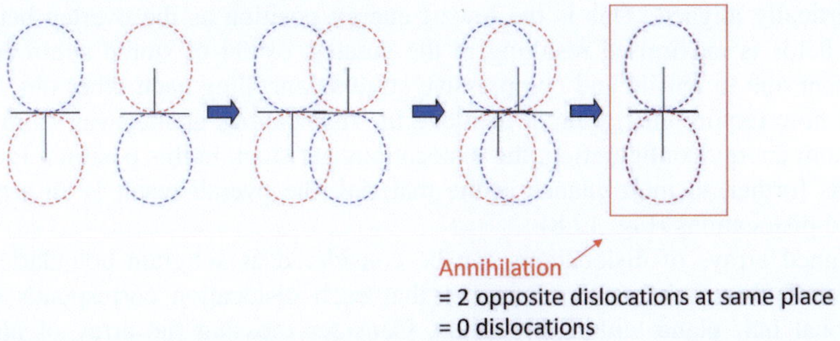

Fig. 12.7 Schematic illustration of the progressive overlap, and cancelling out, of strain fields around two opposite dislocations leading to annihilation. Regions of compressive strain are outlined in red, tensile strain regions in blue

the overlap between the two dislocations, the greater the associated reduction in stored strain energy. This means the two opposite dislocations are attracted to each other. Ultimately, the greatest decrease in stored strain energy occurs when the two opposite dislocations are directly on top of each other. Here the overlap is maximised and the compressive and tensile regions cancel out. It can also be noted that having two opposite dislocations in the same place essentially adds in two extra half planes into the crystal lattice, one above the plane of motion, one below. This is equivalent to simply adding in one additional lattice plane. When the dislocations completely overlap in this way, they cancel each other out. There is an overall decrease in dislocation density as two dislocations have interacted and annihilated each other, transforming into an extra lattice plane and decreasing the total number of dislocations by two.

The terms cold worked and hot worked are used to describe plastic deformation processes for metallic materials. Cold worked materials have a high dislocation density due to the plastic deformation. Hot worked materials have a lower dislocation density as there has been sufficient thermal activation energy to allow recovery and recrystallisation to occur. This counteracts the formation of any dislocations during deformation, decreasing the dislocation density and yield strength, and restoring the ductility of the material.

Annihilation is one of the processes of thermally activated dislocation rearrangement that can occur in cold worked material in order to decrease stored strain energy. Dislocation alignment is a related process, along with annihilation this is part of "recovery", thermally activated dislocation motion driven by a reduction in stored strain energy which allows recovery of original material properties following plastic deformation.

As dislocations of the same type, that is same orientation, but on different lattice planes move through the material they can interact with each other. These interactions are again based on how the strain fields around the dislocations interact. As one dislocation moves past another on a lower plane the compressive strain region of one dislocation interacts with, and partly cancels out, the tensile strain region of the other. In the geometry considered in Fig. 12.8 the dislocations are constrained to move on fixed horizontal planes, the point of maximum overlap is when they are vertically aligned. This is the lowest energy position as the overlap between strain fields is maximised resulting in the greatest extent of stored strain energy reduction due to tensile and compressive strains cancelling each other out. Since it will now require energy input to move the dislocations apart, away from their minimum energy configuration, the dislocations get stuck in this position. Extending this further, to more planes in the material, the overall result is an array of aligned dislocations (Fig. 12.8).

Aligned arrays of dislocations can be considered as subgrain boundaries. To understand this, it helps to remember that each dislocation corresponds to an additional half plane within the lattice. Consider crossing the array of aligned dislocations shown in Fig. 12.9, crossing near the base of the array, just above the first dislocation corresponds to crossing one additional half plane in the lattice. If the array is crossed higher up, say just above the second dislocation then two

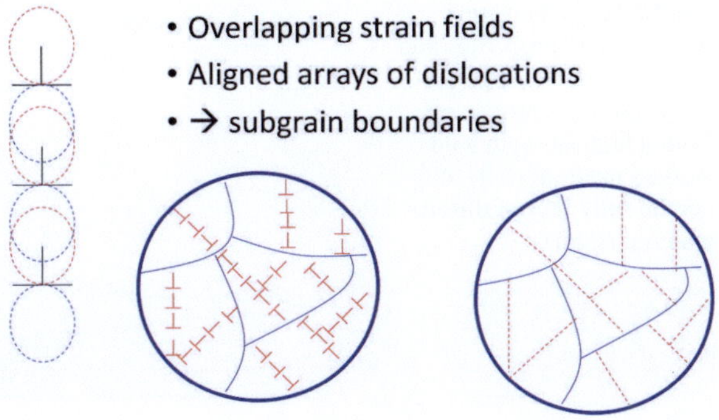

Fig. 12.8 Schematic showing how aligned arrays of dislocations form subgrain boundaries

additional half planes are crossed, that from the first dislocation plus that from the second. This continues up the array, with each additional dislocation adding another half plane to the lattice. Another way of looking at this is that there is a wedge of additional material inserted into the structure at the site of an aligned dislocation array. This is why the arrays can be regarded as subgrain boundaries, the wedge of extra material results in a slight rotation in the orientation of the crystal structure on either side of the array (Fig. 12.9).

During recovery there is a small reduction in dislocation density and an associated small change in material properties. More significant changes to dislocation density and material properties occur during the recrystallisation process.

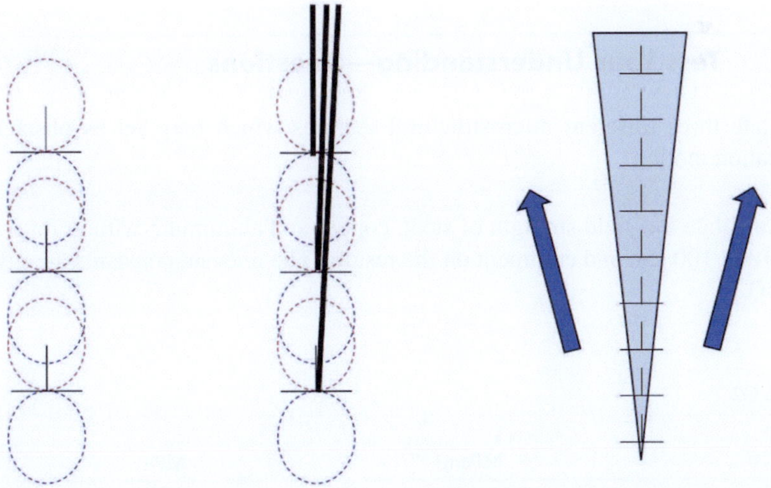

Fig. 12.9 Schematic diagrams showing how the extra half lattice plane associated with each dislocation in an array results in an extra wedge of material which produces a change in orientation of the crystal structure each side of the array, which is why these arrays are regarded as subgrain boundaries

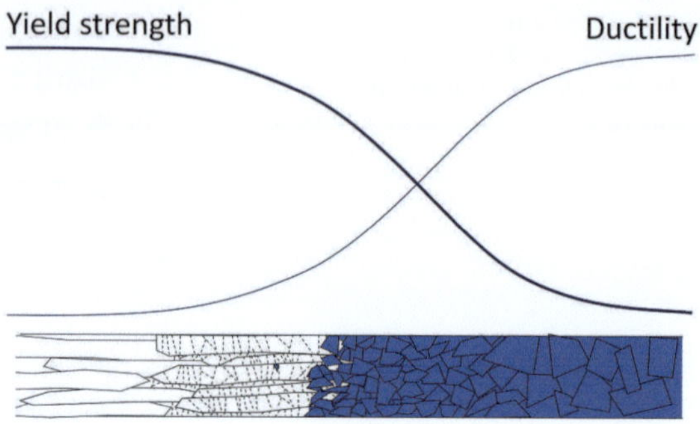

Fig. 12.10 Schematic showing how properties and microstructure change as recrytallisation progresses from a high strength cold worked material (left) to a ductile fully recrystallised material (right)

At the end of recovery there is still a high dislocation density within the subgrains. There will be some variation of dislocation density from subgrain to subgrain. During recrystallisation subgrains with lower dislocation densities grow at the expense of subgrains with higher dislocation densities. As the subgrain boundaries move through the material dislocations are eliminated. Recrystallisation significantly lowers the dislocation density and has a correspondingly large effect on material properties. The yield strength decreases but ductility increases (Fig. 12.10).

If the recrystallisation process is allowed to continue grain growth occurs, further decreasing yield strength and increasing ductility. Recrystallisation can be made use of in manufacturing processes to enable further plastic deformation to be carried out more easily.

12.5 Test Your Understanding—Questions

Q1 State three different microstructural features which may act as obstacles to dislocation motion.

Q2 Calculate the yield strength of steel, copper and aluminium with grain sizes of 10, 50 and 100 μm and comment on the results. The relevant constants are given in Table Q2.

Table Q2

	k MPa $m^{-1/2}$	σ_o MPa
Steel	0.8	80
Copper	0.4	25
Aluminium	0.1	10

Q3 The dislocations shown in figure Q3 are constrained to move horizontally on the lattice planes shown. Determine which dislocations will undergo annihilation and sketch an arrangement of the remaining dislocations that will minimise the total stored elastic strain energy.

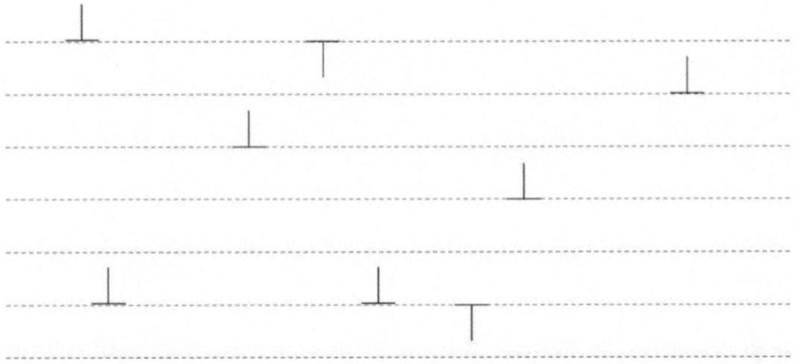

Figure Q3 Schematic distribution of dislocations constrained to move horizontally on the lattice planes shown

12.6 Test Your Understanding—Answers

Q1 Any three of: other dislocations, precipitates, grain boundaries, interfaces, solid solution alloying elements.

Q2 The Hall Petch equation needs to be used

$$\sigma_y = \sigma_o + kd^{-1/2}$$

Steel 10 μm,

$$\sigma_y = 80 + 0.8 \times \left(10 \times 10^{-6}\right)^{-1/2}$$

	k MPa $m^{-1/2}$	σ_o MPa	10	50	100
Steel	0.8	80	333	193	160
Copper	0.4	25	151	82	65
Aluminium	0.1	10	42	24	20

For each material the expected decrease of yield strength with increasing grain size is seen, the material dependency is also evident.

Q3 Answer

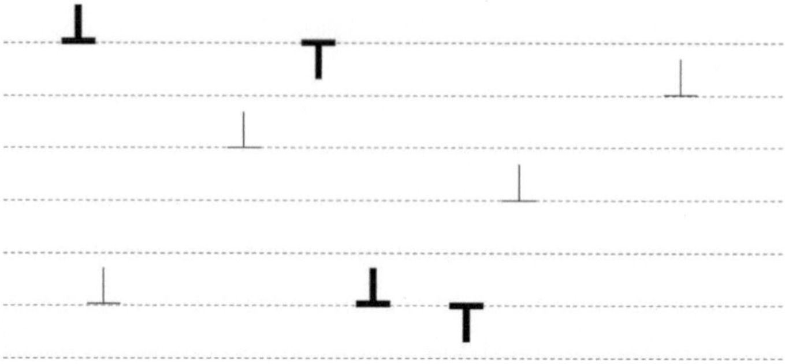

Figure Q3—answer 1

The four dislocations indicated in Figure Q3—answer 1 in **bold** will be removed by annihilation, leaving only the dislocations in Figure Q3—answer 2.

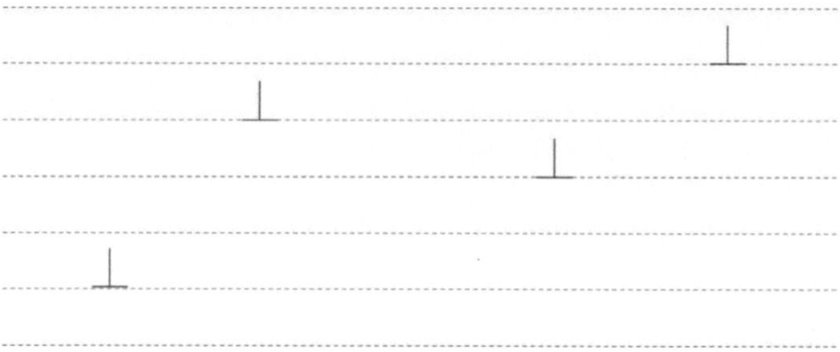

Figure Q3—answer 2

The remaining four dislocations will become arranged in a vertical array as shown in Figure Q3—answer 3, the horizontal position of the aligned array can be anywhere on the figure, the dislocation on the lowest lattice plane should be included in the array as there will still be some, albeit less, overlap between it and the dislocation 2 planes above it.

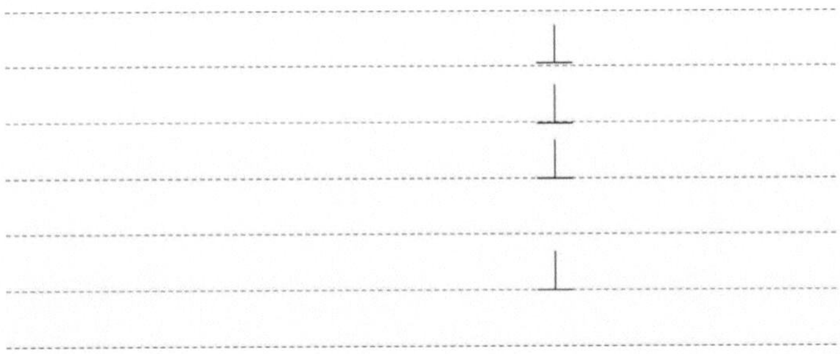

Figure Q3—answer 3

Further Reading

There are a number of tutorials related to dislocations in https://www.doitpoms.ac.uk/index.php

There are also many websites with information on dislocations, that of The University of Kiel is particularly useful as it includes an animation of dislocation motion https://www.tf.uni-kiel.de/matwis/amat/iss/kap_5/backbone/r5_4_1.html

Reference

1. https://journals.aps.org/prl/abstract/10.1103/PhysRevLett.127.100401

Key Concepts and Glossary

Specific properties The term specific properties does not mean particular, specified properties, but rather properties that have been divided by the density. Specific strength is strength divided by density and specific stiffness is stiffness divided by density. It is useful to consider specific strength and specific stiffness as this is what determines how much of the material is required to meet the design constraints. Component mass scales with these specific properties. Specific properties are particularly relevant to transport industries where mass minimisation is a priority.

Thermodynamic driving force Thermodynamics dictates that systems move towards states which minimise their energy. When a change is described as having a thermodynamic driving force it means that after the change has been implemented the energy of the system is decreased, hence the change is thermodynamically favoured.

Lightweighting This term is widely used, particularly in the transport industries, to refer to the deliberate decrease in mass of components and vehicles. This is driven by the wish to minimise the energy required to move the vehicle or component and the associated power and energy requirement savings.

Hotter is better This idea underlies a lot of engine material development. Higher engine operating temperatures are desirable due to thermodynamic reasons which mean that increased efficiencies can be obtained at higher operating temperatures.

"Grain" is the term used in materials to refer to individual crystals in a polycrystalline material.

"Grain boundaries" are the interfaces between adjacent grains in a structure. These are important as they are regions of disorder which act as obstacles to dislocation motion.

"Lattice" refers to the crystal structure, with lattice planes being different planes within the structure.

"Microstructure" is the inner structure of a material. It can be thought of as what is seen when a material is examined with a microscope. A material's microstructure is closely related to its material properties, with material properties being affected

K. T. Voisey, *The Engineer's Guide to Materials*,
https://doi.org/10.1007/978-3-031-62937-2

by microstructural features. Microstructural features typically tend to range in size from sub micron to hundreds of micrometers.

"Phase" is a word which can cause confusion by the way it is used in materials. In materials a phase is a physically distinct form of matter. Yes, gases, liquids and solids (and plasmas) are all different forms of matter but the materials usage of the word goes further than this. You can have a two phase material where both phases are solid. In AA2024, the copper rich precipitates have a different composition to the surrounding aluminium. This means that there are two physically distinct forms of material present, making this a two phase material, even though both phases are solid.

"Precipitates" are particles of a different phase embedded in the main crystal structure. These can cause a significant disruption to the crystal lattice.

"Solution" is another word used in a slightly different way in materials compared to its general usage. A solution is one material dissolved in another, however in materials this can be in the solid state with an alloying element dispersed throughout a solid material, such as carbon in low carbon steel. This is also referred to as a "solid solution".

Index

K. T. Voisey, *The Engineer's Guide to Materials*,
https://doi.org/10.1007/978-3-031-62937-2

D

T

MIX
Papier aus verantwortungsvollen Quellen
Paper from responsible sources
FSC® C105338
FSC
www.fsc.org

If you have any concerns about our products,
you can contact us on
ProductSafety@springernature.com

In case Publisher is established outside the EU,
the EU authorized representative is:
Springer Nature Customer Service Center GmbH
Europaplatz 3, 69115 Heidelberg, Germany

Printed by Libri Plureos GmbH
in Hamburg, Germany